01-HHE-21T

TOWARDS AN INNOVATIVE DNA ARRAY TECHNOLOGY FOR DETECTION OF PHARMACEUTICALS IN RECLAIMED WATER

by:

Seth W. Kullman
Karl G. Linden
David E. Hinton
Duke University

2007

WERF

Awwa Research Foundation
Advancing the Science of Water®

NWRI

WATEREUSE

California Urban Water Agencies

The Water Environment Research Foundation, a not-for-profit organization, funds and manages water quality research for its subscribers through a diverse public-private partnership between municipal utilities, corporations, academia, industry, and the federal government. WERF subscribers include municipal and regional water and wastewater utilities, industrial corporations, environmental engineering firms, and others that share a commitment to cost-effective water quality solutions. WERF is dedicated to advancing science and technology addressing water quality issues as they impact water resources, the atmosphere, the lands, and quality of life.

For more information, contact:
Water Environment Research Foundation
635 Slaters Lane, Suite 300
Alexandria, VA 22314-1177
Tel: (703) 684-2470 Fax: (703) 299-0742
www.werf.org werf@werf.org

This report was co-published by the following organization. For nonsubscriber sales information, contact:

IWA Publishing
Alliance House, 12 Caxton Street
London SW1H 0QS, United Kingdom
Tel: +44 (0) 20 7654 5500
Fax: +44 (0) 20 7654 5555
www.iwapublishing.com
publications@iwap.co.uk

Library of Congress Catalog Card Number: 2005921994

IWAP ISBN: 978-1-84339-737-3

Duke University

The research on which this report is based was developed, in part, by the United States Environmental Protection Agency (EPA) through Cooperative Agreement No. CR827345-01 with the Water Environment Research Foundation (WERF). However, the views expressed in this document are solely those of Duke University and neither EPA, WERF, AwwaRF, National Water Research Institute, WateReuse Foundation, nor California Urban Water Agencies endorses any products or commercial services mentioned in this publication. This report is a publication of WERF, not EPA. Funds awarded under the Cooperative Agreement cited above were not used for editorial services, reproduction, printing, or distribution.

ACKNOWLEDGMENTS

Report Preparation

Principal Investigator(s):

David E. Hinton, Ph.D.
Seth W. Kullman, Ph.D.
Karl G. Linden, Ph.D.
Duke University

Project Team:

David C. Bencic, Ph.D.
David E. Hinton, Ph.D.
Seth W. Kullman, Ph.D.
Karl G. Linden, Ph.D.
David C. Volz, MS
Duke University

Project Subcommittee

Robert Berger
Municipal Utility District (retired)

Louis J. Guillette Jr., Ph.D., PSC chair
University of Florida

Mary E. Buzby, Ph.D.
Merck & Company, Inc.

Gary R. Klinefelter, Ph.D.
U.S. Environmental Protection Agency

Jerry Ongerth, Ph.D.
East Bay Municipal Utility District

Water Environment Research Foundation Staff

Director of Research: Daniel Woltering, Ph.D.
Program Director: Jami Montgomery
Program Manager: Jane M. Casteline

ABSTRACT AND BENEFITS

Abstract:

Numerous pharmaceutical substances and their metabolites have been identified in the aquatic environment. Recent interest in quantifying pharmaceuticals in waste effluents and aquatic environments has identified the need to develop bioanalytical assays that will accurately reflect presence of these compounds. The goal of this research was to investigate the use of DNA microarrays as a bioanalytical tool for analysis of pharmaceuticals in treated waters that are intended for subsequent further use, i.e. reuse waters. Results from this study suggest that prototypic pharmaceuticals target molecular events associated with pharmaceutical-induced gene induction and repression in exposed organisms. Gene expression profiles are subsequently evaluated as a means to discern pharmaceutical exposure. When fully developed, such a bioanalytical approach should provide an efficient and robust method for screening pharmaceuticals in reuse and other water matrices.

Benefits:

- Identifies new environmental biomarkers by analysis of induction and/or repression of select gene markers and recognition of specific gene expression signatures.
- Establishes validity of using DNA arrays to rapidly screen changes in gene expression in response to aquatic pharmaceutical exposure.
- Improves environmental monitoring and accelerates the ability to rapidly assess chemical exposure.
- Screens thousands of genes simultaneously eliminating the time and cost of traditional biomarker analysis.

Keywords: Microarray, medaka fish, pharmaceuticals, bio-analytical analysis, reuse water.

TABLE OF CONTENTS

LIST OF TABLES

LIST OF FIGURES

LIST OF ACRONYMS

AMP	ampacillin
APR	acute phase response
Array	macro/microarray
CDNA	complimentry DNA
CHG	choriogenin
CPM	counts per minute
CWA	Clean Water Act
CYP	cytochrome
dCTP	deoxycytidine triphosphate
DNA	deoxyribonucleic acid
dNTP	dinucleotide triphosphate
DSBR	Duke Bioinformatics Shared Resource
DTT	dithiothreitol
EDC(s)	endocrine disrupting compounds
EDTA	ethylenediaminetetraacetic acid
EST-	expression sequence tag
FDA	Food and Drug Administration
g	gram
GC/MS	gas chromatography/mass spectroscopy
GCG	Genetics Computer Group
IACUC	Institutional Animal Care & Use Committee,
kDa	kilodalton
Kg	kilogram
L	liter
LB	lauria broth
LC/MS	liquid chromatography/mass spectroscopy
LC/MS/MS	liquid chromatography/mass spectroscopy/MS
mg or ml	milligram or mililiter
mRNA	messenger RNA
NCBI	National Center for Biotechnology Information
ng	nonagram
NIEHS	National Institute of Environmental Health
NPDES	National Pollutant Discharge Elimination System
NR	Nuclear receptor
NSAID	Non steroidal anti inflammatory drugs
OEHS	Office of Environmental Health and Safety
PCR P	olymerase Chain Reaction
PMSF	Phenylmethylsulphonylfluoride
PPAR	Peroxisome proliferator-activated receptor
RNA	ribonucleic acid
rRNA	ribosomal ribonucleic acid
RXR	retinoid X receptor
SSC	sodium citrate buffer
SSH	suppressive subtractive hybridization

TE	tris EDTA
tRNA	transfer ribonucleic acid
μg or μl	microgram or microliter
U.S. EPA	United States Environmental Protection Agency
U.S.G.S.	United States Geological Society
UV	Ultraviolet
VTG	Vitellogenin
WET	whole effluent toxicity testing

EXECUTIVE SUMMARY

U.S. surface waters are subjected to increasing anthropogenic stress coupled with natural stressors such as: salinity, dissolved oxygen, temperature, nutrient inputs and physical energy. River and coastal waters/sediments are the ultimate recipients of an array of pollutants released in municipal and industrial point discharges, and carried in urban and agricultural non-point runoff. River and coastal regions throughout the U.S. have been impacted by chemical stressors including pesticides, PAHs and PCBs. More recently, attention has been directed towards the potentially hazardous effects of Pharmaceuticals and Personal Care Products (PPCPs) and hormonally active substances. Reports of pharmaceutical usage often are in the range of tons of active substance/per year with defined daily doses in the millions (Halling-Sorensen et al., 1998). Based on estimates of prescription numbers and usage patterns, it has been demonstrated that a diversity of pharmaceuticals are ultimately deposited in aquatic receiving environments. These and other studies have additionally indicated that numerous pharmaceutical compounds are recalcitrant to current waste treatment processes and present in low but consistent concentrations in receiving waters (Daughton and Ternes, 1999).

With unique chemical and physical characteristics, analytical procedures for analysis of pharmaceuticals in water matrices has proven difficult. This is due in part to due to the large numbers of compounds, the low concentrations in which they are found in the environment and inability to develop broad extraction procedures. To circumvent these issues, biological methods are becoming increasingly popular as broad based environmental screening tools. Successful bioassays have been designed to screen for some pharmaceuticals such as estrogenic or androgenic environmental endocrine disruptors (EDCs), however these represent a small subset of pharmaceuticals and to date, bioassays for most pharmaceutical compounds have not been developed.

In this study, researchers have used recent advances in toxicogenomics, a sub-discipline derived from the combination of genomics and traditional toxicology, to develop signature biomarkers of pharmaceutical exposure in treated waters that are intended for subsequent further use, i.e. reuse waters. The impact of genomic technology on the progress of biomedical and environmental science is difficult to overemphasize. The potential rewards of the multiple species genome projects in providing detailed blueprints of genomes are enormous. Already, this information is accelerating the pace of gene discovery and technologies to measure gene expression on a global scale. The development of DNA microarray methodology that enables the simultaneous analysis of tens of thousands of genes has transformed research that aims to understand gene regulation and pathways of gene control. This technology provides the means to assess global changes in gene expression in response to environmental contaminants. Further, changes can be explored without the need for prior knowledge or bias in the targets that are measured. The ability to measure large-scale gene expression represents a unique tool for defining exposures of cells or organisms. The pattern of genes expressed following exposure to environmental toxicant can be used as a signature of that toxicant (i.e., a biomarker of toxicant exposure).

The general premise of this research is that pharmaceuticals, at environmentally relevant concentrations will elicit molecular changes in exposed organisms that can be quantified and employed as a bioanalytical tool for exposure assessment. In this study a bioassay was developed that targets molecular events associated with pharmaceutical-induced gene induction and repression in exposed surrogate organisms. The objective is to use pharmaceutical-induced gene changes as a means to assess exposure to and thus their presence in aquatic environments. This bioassay approach permits evaluation of biological effects of pharmaceuticals and significantly enhances ongoing efforts to understand and address the short and long-term consequences of these compounds in the environment.

These studies were initiated by developing a library of gene changes (differential expression libraries) in medaka fish following exposure to a prototypic class of pharmaceuticals found in aquatic environments; antihyperlipidemics (lipid lowering drugs). This class was chosen based on their identification as emerging contaminants by the U.S.G.S. Toxic Substances Hydrology Program and their well characterized mechanism of action enabling proof of principle for this bioassay approach. Differential expression libraries were produced by suppressive subtractive hybridization (SSH) (Diatchenko et al., 1996). Subtractive hybridization enables the comparison of two populations of genes that are expressed in one population but not the other i.e. control vs. treated. SSH is a highly effective and efficient method to enrich for differentially expressed rare genes. Using this approach, approximately 1000 individual genes were randomly isolated and characterized from multiple medaka tissues. All genes were collected and sequenced to establish an annotated gene library. When compared, 75% of the genes exhibited known gene homologies to other species. Of these, 60% of sequences were non-redundant representing ~450 different genes. Subtraction libraries demonstrated that differential gene expression was highly tissue specific with few genes expressed in multiple tissues.

To facilitate interpretation of results, genes were grouped according to their cellular processes (gene ontology) in which they are likely to be involved. This information is generated

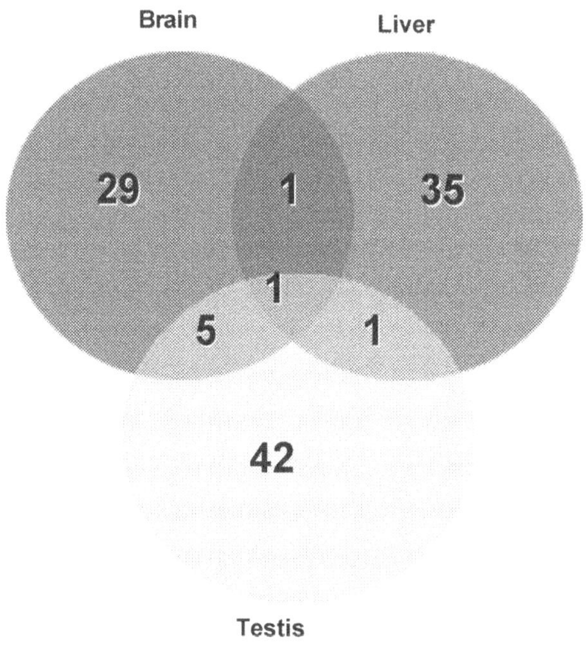

Figure ES-1. Venn Diagram Illustrating Numerical Distribution of Top 100 Non-redundant Genes Form Medaka Tissues Following 50mg/kg Injected Adult Male Medaka.

through public data bases including, Pub Med, Swiss-Prot, LocusLink, and GO annotation databases. Genes were annotated with processes that are consistent with the pharmacological effects of fibrate drugs in mammalian studies. Even though many genes may be involved in more than one process, each gene was grouped into only one category: including genes involved in the metabolism and transport of lipids, glucose homeostasis, amino acid metabolism, cell proliferation and differentiation, and apoptosis and components of the immune and inflammation response.

Subsequent studies were focused on the development of microarrays for analysis of fibrate pharmaceuticals using the medaka fish. The general approach taken was to apply the gene sets acquired by SSH to membrane based arrays and assess gene expression as markers of pharmaceutical exposure. The goal of this approach was to develop a set of gene markers that can be applied to identify the presence of fibrate based pharmaceuticals in reuse waters. First it was necessary to establish specific parameters associated with array fabrication and dynamics including spotting of DNA targets, non-amplified versus amplified probe production, software for image analyses, and development of internal standards for normalization. Each of these factors proved critical for quality control of microarrays and has contributed significantly to the consistency of our analysis. Once arrays were established, a multi-factorial experimental design that incorporated dose, time, tissue and compound specific effects was conducted. Analysis of gene expression was initially analyzed using unsupervised evaluations including hierarchical clustering and principal component analysis for determination of gene expression trends

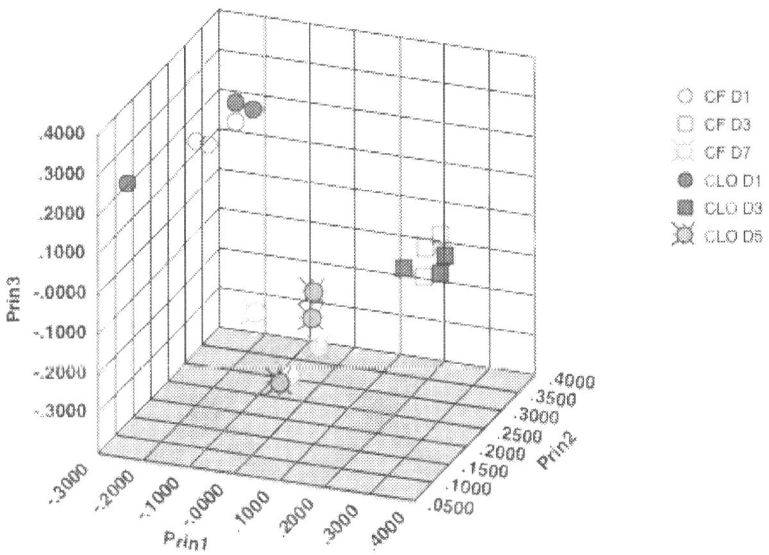

Figure ES-2. Principal Component Analysis. Ciprofibrate and Clofibrate Days 1, 3, and 7.

followed by compression of the datasets to identify specific expression profiles and biomarkers. Statistical analysis was performed on all data sets to identify significant gene responses using a mixed function linear model with a Bonferonni adjustment (Wolfinger et al., 2001). As array data were generated using a nested experimental design, a mixed model was used to account for variability linked with each level of experimental design.

Microarray data analysis demonstrated that gene expression patterns reveal several key events following treatment with fibrate based antihyperlipidemics including time dependent up-regulation, time dependent down regulation and transient up or down regulation. Time to response was the major contributing factor to the global organization of the dataset. Secondarily, dose response relationships were evident on a per day basis. Using a statistical permutation process, datasets were examined for linear trends between each dose. Four distinct linear trends were observed: in the first linear trend, typical up regulation was observed as a consistent increase in gene expression form control to low dose and upward to the high dose. Secondly, linear trends were identified in numerous genes as a consistent down regulation. Non-linear dose response changes were additionally observed with both liver datasets. These gene expression relationships were more indicative of a non-monotonic, biphasic "U" or inverted "U" shaped dose response where the middle dose is distinctly different from the lower and higher dose. This type of dose response is often observed with endocrine disrupting compounds and is a subject of study. From a biomarker perspective we are more interested in genes that are consistently up or down regulated following exposure to the compound of interest. Ideally these gene sets demonstrate a dose response relationship, and thus an estimation of exposure can be determined based on the degree of expression. Alternatively, a qualitative determination of exposure can be made based on observation of a specific gene expression profile. In keeping with this design we have compared gene expression response between both ciprofibrate and clofibrate for gene sets that are either consistently up or down regulated across dose.

Given that two fibrate-based pharmaceutical compounds that share a similar mechanism of action (i.e. nuclear receptor-PPARα agonists) were used in this study, it was hypothesized that similar gene expression profiles would be observed. This hypothesis is the basis for the identification of chemical class specific markers that are used to discern exposure to select compounds. The assumptions to this approach however are that both compounds display similar binding kinetics to the PPAR receptor (i.e. similar Km's) and if any, share similar non-receptor mediated mechanisms/effects. Our comparison of gene expression across both dose and time resulted in a highly consistent pattern of expression between the two pharmaceuticals. Given the degree of consistency in gene expression it is anticipate that fibrate compounds will elicit a common set of gene markers in medaka that can be exploited biomarkers of exposure. As an additional level of scrutiny the researchers have applied a stringent requirement that individual genes changes in treated animals are statistically significant from the controls. For expression profiling and biomarker analysis, it is important to demonstrate that gene expression is not only up or down regulated but that gene expression responses are statistically significant. The mixed model provides assurance that gene changes accurately reflect molecular events associated with exposure rather than a random event or type one statistical errors. While the initial analysis was based on gene expression changes between 1.3 fold up and 0.75 fold down, by applying robust statistical tests the stringency for identifying a truly differentially regulated gene sets is increased. This analysis results in two specific changes to the expression profile; first,compression of the data set by elimination of genes that while highly up or down regulated were not statistically significant and second, expansion of the data set by addition of genes with high significance but not highly up or down regulated.

Three general requirements were used to identify a gene set to evaluate fibrate-class specific exposures: 1) high to moderate gene expression change from control value; 2) statistically significant difference between control and treated gene expression values; and 3) dose response relationships. One caveat to this type of analysis however is an assumption that other classes of pharmaceuticals do not produce a similar gene expression profile. Thus a comparison was made between gene expression changes between the fibrates (ciprofibrate and clofibrate) and two prototypic estrogens (17β estradiol and ethinyl estradiol) representing two distinct pharmaceutical classes. Results of this analysis suggest that global gene expression profiles are highly unique for each pharmaceutical class. However, with further examination of individual genes, overlaps in gene expression were noted. This observation further supports the need to evaluate global expression profiles rather than rely solely on one or two specific genes.

While these data sets met most of these criteria, there is room for improvement. The researchers envision this first set of arrays as development of a training dataset (i.e. the gene expression profile), which can be subsequently improved upon and challenged with increasing repetition. In subsequent research, data should be acquired from exposures at lower concentrations, where both time and dose will drive global organization of the gene expression response and relate gene expression to environmentally relevant concentrations. Secondly, increasing both the biological and technical replication of the arrays will aid in generating more statistical relevance, validating gene expression changes. Third, additional members of the pharmaceutical class should be compared to the original training set to demonstrate consistency in gene expression response and last analysis between pharmaceutical classes should be conducted to identify distinct differences in gene expression responses.

As a last component of the research plan, microarray studies were performed with a reuse water from a local treatment facility. These studies encompassed analysis of reuse water alone and reuse water spiked with two low doses of clofibrate for 24 and 72 hours. Results from this study demonstrate that the microarray array procedure and analysis provides robust and statistically significant process for establishing compound specific gene expression changes in reuse water. Interestingly, when compared to laboratory water, the reuse water alone resulted in large number of statistically significant gene changes. The genes included several candidate markers for fibrates in addition to gene previously identified as responsive to estrogens and aromatic hydrocarbons (other studies in the researchers laboratory), providing an initial indication that reuse water used for this study contained a mixture of urban contaminants at concentrations capable of eliciting gene expression changes in our model system. Analysis of reuse water spiked with cholfibrate suggests that gene regulation occurred in both an up and down direction. Gene expression changes were most pronounced on day one in the low dose treatment and in some instances a dose response effect was observed. Fewer statistically significant genes changes were observed on day three. This is likely due to a prolonged exposure to potential reuse water conditions resulting in a high degree of background in the control samples. Thus, results suggest that array analysis is sensitive and specific for identification of gene expression changes in reuse water during short exposure durations.

In the most likely scenario, reuse waters with unknown contaminants will be compared to laboratory waters where background expression values will not interfere with analysis. With further development of arrays and screening of prototypic contaminant classes, training sets can be established to further develop class prediction models for the identification of compounds, or compound classes from complex mixtures of water contaminants. A key to this approach will be the ability to clearly identify non-overlapping gene markers that comprise an established gene expression profiles. This approach will enable a focus on the patterns of gene expression or expression profiles and observe distinct differences in gene regulation. Additionally, if gene annotations are examined, it becomes obvious that specific chemical classes will elicit defined metabolic pathways based upon their mechanism of action. As described in chapter 2.0, fibrate drugs are PPARα agonist and thus result in expression of a defined set of genes and pathways associated with lipid metabolism and transport, amino acid metabolism and others that can be exploited for profile analysis

Through this analysis, the researchers have generated a specific set of gene expression profiles useful as a bioanalytical tool for chemical assessment in reuse waters and other aquatic environments. As with any new tool it is believed that use of microarray analysis will take time to become an established assessment method. These studies were designed as proof of principle for such an application. Given these results and the rapid growth in genomics technologies, the water quality community is poised to evaluate the feasibility of array analysis for assessment of select pharmaceuticals in reuse and other water matrices. This approach will greatly aid in assessment and analysis of potential environmental and human risk in the near future. A work flow incorporating microarray array analysis could be developed for other pharmaceutical compounds of interest in reuse water following the procedure used in this research; that is,

- Identify specific compounds of interest in reuse water by examination of urban population pharmaceutical use patterns and demographics.
- Develop gene expression training sets by screening of prototypic compound classes, in species of interest (in this case the medaka fish) with established microarrays.
- Ascertain dose response effects.
- Ascertain time to response effects.
- Clearly determine non-overlapping gene markers that comprise an established gene expression profile.
- Validate gene expression events by analysis of biochemical pathways based upon known mechanisms of pharmaceutical action.
- Determine effects of reuse water matrix on gene expression (TOC, alkalinity, DO etc.)
- Establish a smaller set of gene expression changes that are indicative of chemical exposure.
- Once established, conduct real time polymerase chain reaction (PCR) in place of microarray studies. This analysis will be better suited to the technical capabilities of interested utilities.

CHAPTER 1.0

INTRODUCTION, RATIONALE AND PILOT STUDIES FOR MICROARRAY DEVELOPMENT

1.1 Introduction

This chapter provides a background on the approaches used to establish a medaka specific microarray for analysis of aquatic pharmaceuticals in reuse water. The aim of this chapter is to outline a detailed description of the rational used for the study and guide the reader on the over all series of events conducted to produce the array. Since at this time there are no commercially available microarray platforms for medaka, these studies were designed to generate a targeted array that would be useful for bioanalytical assessment of fibrate class of antihyperlipodemic pharmaceuticals. These arrays were established as a proof of principle and studies were conducted to demonstrate the utility of this approach for identifying select gene markers for determining exposure to and thus presence of these compounds. Targeted arrays are designed specifically with a known set of gene probes predicted to be responsive to the desired treatment. These arrays differ from genomic arrays or open architecture arrays where all known genes (ESTs, cloned cDNAs or oligonucleotide sequences) are placed on an array. We have chosen to make targeted arrays based on the ability to identify "fibrate responsive gene sets" using subtractive hybridization.

1.2 Rationale

1.2.1 Overview, Aquatic Toxicity Testing

The Federal Water Pollution Control Act, commonly known as the Clean Water Act (CWA), was enacted in 1972 with the objective of "restoring the chemical, physical and biological integrity of the Nation's waters." Several goals and policies were established in the act including eliminating the discharge of pollutants and maintaining water quality sufficient for the protection and propagation of fish, shellfish, and wildlife. In 1987, the CWA stated that "no toxics in toxic amounts" shall be discharged into U.S. waters (U.S. EPA, 1991). In the 29 years since the CWA was enacted, the U.S. Environmental Protection Agency (U.S. EPA), through the National Pollutant Discharge Elimination System (NPDES) permitting program, has made significant progress toward controlling toxic discharges and implementing water quality standards. Industries and municipalities must comply with these federal standards by conducting whole effluent toxicity testing (WET) using a three species assays (U.S. EPA, 1993a; 1993b; 2000). Although WET testing has led to a vast improvement of surface water quality, the effort is directed at acute toxicity. Endemic organisms however, are now exhibiting chronic toxicities including developmental impairment, altered sexual differentiation, and impaired reproduction from exposure to compounds undetected by these analysis including numerous endocrine disruption chemicals and pharmaceutical agents. To address this issue the U.S. EPA has issued mandates initiating testing programs to determine which chemicals act to disrupt the endocrine system. The Food Quality Protection Act (1996) and the Safe Drinking Water Act (1996) and an

Endocrine Disruptor Screening and Testing Advisory Committee (EDSTAC) has recommended use of cell assays and QSAR to *screen* industrial and agricultural chemicals for endocrine-disrupting effects (Ankely et al., 1998). Largely unaddressed however is the potential impact of aquatic pharmaceuticals on human and environmental health. Few assays have been developed to ascertain the presence and/or effects of these compounds in surface and/or reclaimed waters.

1.2.2 Pharmaceutical Fate Distribution and Detection in the Aquatic Environment

U.S. surface waters are subjected to increasing anthropogenic stress coupled with natural stressors such as: salinity, dissolved oxygen, temperature, nutrient inputs and physical energy. River and coastal waters/sediments are the ultimate recipients of an array of pollutants released in municipal and industrial point discharges, and carried in urban and agricultural non-point runoff. River and coastal regions throughout the US have been impacted by chemical stressors including pesticides, PAHs and PCBs. More recently, attention has been directed towards the potentially hazardous effects of Pharmaceuticals and Personal Care Products (PPCPs) and hormonally active substances. Reports of pharmaceutical usage often are in the range of tons of active substance/per year with defined daily doses in the millions (Halling-Sorensen et al.,, 1998). Based on estimates of prescription numbers and usage patterns, it has been demonstrated that a diversity of pharmaceuticals are ultimately deposited in aquatic receiving environments (Sedlak, 2000). These and other studies have additionally indicated that numerous pharmaceutical compounds are recalcitrant to current waste treatment processes and present in low but consistent concentrations in receiving waters (Daughton and Ternes 1999). Studies in the river Lee, England, demonstrated the presence of over 170 different pharmaceutical compounds with concentrations exceeding 100 ng/l (Richardson and Bowron, 1995). This situation is additionally exacerbated by the use of veterinary pharmaceuticals, which, following animal excretion are often deposited directly into streams and other aquatic environments. Thus over the last decade there has been an increasing awareness that environmental fate and distribution of pharmaceuticals has been largely underestimated. With unique chemical and physical characteristics, standard analytical procedures aimed at detecting persistent organic pollutants have proven problematic for detecting the vast array of potential environmental pharmaceutics. Thus an intensive effort has been made to develop procedures to quantify many of these compounds from environmental matrices. In general pharmaceuticals and their metabolites found in aquatic environments have low vapor pressures, low partition coefficients and contain polar functional properties (Sedlak, 2000). As such these compounds are not suitable for typical organic solvent extractions and GC/MS analysis. Solid phase extraction has, however, been successful for concentrating pharmaceuticals for numerous matrices and analysis has been accomplished with the increased analytical sensitivity of LC/MS and LC/MS/MS (Sedlak, 2000). To date however, methods of detection of pharmaceuticals have relied solely on analytical measurements. Few bioassays have been developed to assess the exposure and effect of these compounds on non-target organisms.

Successful bioassays have been designed to screen for endocrine modulating compounds (EDCs) however these represent a small subset of pharmaceuticals. Based on the premise that EDCs operate via similar modes of action between often distantly related species (i.e. estrogen receptor and androgen receptor agonists and antagonists), rapid *in vitro* and *in vivo* assays have been developed to ascertain presence of these compounds in various water matrices and or biological effects. These assays include hormone receptor binding assays, reporter systems to measure transcriptional activation, and the proliferation of MCF-7 breast cancer cells (E-screen) (Garcia-Reyero et al., 2001; Jorgensen et al., 2000; Soto et al., 1995). Additionally since reproductive proteins including vitellogenin (Vtg) and choriogenin (Chg) induction are under the control of estrogen in oviparous organisms, it has been broadly used as a biomarker of exposure for xenoestrogens (Arukwe et al., 2001; Arukwe et al., 1997).

The biomarker approach to biomonitoring in aquatic toxicology has been reasonably successful for determining exposure and effect in feral or surrogate species exposed to select contaminants (Mayer et al., 1992). Based on identification of contaminant-induced variations in molecular, biochemical physiological and/or histological endpoints, this approach aims at correlating condition or "health" of an organism with exposure. Thus the goal of the biomarker approach has been to develop integrative measurements that quantify exposure in a dose or time dependent manner and relates them to the degree of an organism's dysfunction. Biomarkers historically have been applied at many levels of biological organization and are usually categorized into specific and nonspecific responses. However, markers that measure contaminant-induced changes at the molecular level are often more sensitive than those at higher levels of organization and are usually the first detectable and quantifiable responses to environmental change. Thus by exploiting our knowledge of both chemical behavior and mechanisms of toxicity, biomarkers have been developed to ascertain exposure to select classes of environmental toxicants, or have been used as an indication of a more generalized stress response (Ara et al., 1994).

1.2.3 Fate of Chemicals in Wastewater Treatment Processes

Although many pharmaceuticals in wastewater can be degraded biologically, this degradation is not always complete nor are all pharmaceuticals degraded during treatment. Richardson and Bowron (1985) identified many pharmaceuticals such as antibiotics and pain relief medications that were persistent following conventional wastewater treatment. Persistent and partially degraded contaminants must be treated further before reuse (Ternes, 1998). The effects of treatment processes used in reuse facilities such as UV irradiation, oxidation, or chlorine/chloramination may further degrade or possibly increase toxicity of pharmaceuticals in water. For instance, although ozone alone could not degrade clofibric acid or ibuprofen, the ozone/peroxide process which generates highly active hydroxyl radicals improved the degradation efficiency, however toxicity was not investigated (Zwiener and Frimmel, 2000). Although not studied thus far, chloramination of reuse water could alter the toxicity of persistent or partially degraded compounds. Better understanding of how these treatment processes affect toxicity of pharmaceuticals in reuse waters can eventually lead to improved treatment strategies aimed at producing toxicity-free effluents.

1.2.4 Gene Arrays

With the emerging technology in gene expression analysis, biomarker applications in applied toxicology can be greatly improved. Genome expression technologies are becoming increasingly reliable and accessible (Granjeaud et al., 1999). It is now feasible to generate total mRNA expression profiles for whole genomes following stimuli such as exposure to toxicants (Medlin, 1999). The biomedical and pharmaceutical industries have greatly exploited these new technologies to identify new drug targets, delineate mechanisms of drug action, and identify specific prognostic markers for clinical disease (Rockett and Dix., 2000). For many applications, DNA arrays have become the preferred method for large-scale mRNA expression analysis (Granjeaud et al., 1999). Interest in the application of DNA- array analysis for toxicity studies is rapidly growing and has proven useful for defining mechanisms of toxicity and classifying environmental contaminants based on characteristic gene markers and expression profiles (Nuwaysir et al., 1999; Bartosiewicz et al., 2001). Establishment of the Toxicogenomics

Research Consortium initiated by the National Institute of Environmental Health Science (NIEHS) and the development of a gene expression profile database for contaminant exposures through the National Center for Toxicogenomics (NTP) has impressed the importance of DNA-array applications in this regard. Thus a current focus in toxicity studies has been the development of expression profiles for selected environmental contaminants. These profiles are being generated through the use of customized DNA arrays that contain known gene sequences associated with basic cell function and specified targets of toxicity. To date, several arrays have been developed, comprised of gene clusters for apoptosis, cell cycle, metabolism, oncogenes and others (Pennie et al., 2000; Bartosiewicz et al., 2001; Medlin, 1999). The advantage of this DNA-array design is the selective targeting of known responsive genes resulting in a highly customized signature of toxicity. One additional approach is to array thousands of unknown gene sequences from organisms of interest. Following hybridization with cDNA from treated tissue/cells/organisms, positive genes are identified and further characterized (De Benedetti et al.,, 2000). Compared to contemporary standardized bioassays, it is expected that DNA-array assessment of total mRNA expression will greatly improve the accuracy and efficiency of toxicity screening (Pennie et al., 2000). It is believed that this technology will have a broad base of applications pertaining to many aspects of toxicology including assessment of toxicity in drug development, elucidation of mechanism of action, prediction of disease or health status, determining occupational exposure and monitoring pollutant exposure in the environment.

To date, application of DNA-array technology for aquatic environmental monitoring has received little attention. This is most likely due to the fact that commercial arrays are not yet available for most fish models (Affymetrix and Agilent recently made a zebrafish array available). Small aquarium fish species including medaka are increasingly being used as vertebrate models in various fields of biology including toxicology, genetics, and developmental biology. These organisms are suitable for use in genomic studies due to their small genome size, and ease of application in forward genetics (Ishikawa, 2000). Similarly, medaka fish have become a standard model organism in aquatic toxicity studies (Papoulias et al., 1995; Patyna et al., 1999). Their transparent chorions, reproductive capacity, external gender characteristics, and sensitivity to toxicants are but a few of the characteristics that have made them an excellent model for toxicity studies. A genome project has been initiated for medaka and databases containing expression sequence tags (EST) and genomic DNA sequences have been established. The availability of coding sequence information has enhanced the ability to directly target specific genes into an array. An additional approach is to identify target genes by differential gene expression. In this research, we targeted candidate genes for DNA-array construction by first developing a pharmaceutical induced expression library using subtractive hybridization. Identified genes were subsequently arrayed and used to characterize pharmaceutical exposures. Through this process we have developed a highly customized DNA-array comprised of specific pharmaceutical responsive genes applicable for environmental monitoring purposes.

Why aquatic gene DNA arrays? Were gene arrays available, monitoring biological effects would further our ability to determine deterioration in environmental quality, maintain fish and wildlife resources and assess potential risk to public and ecological health. Application of gene expression profiling will greatly improve environmental monitoring and accelerate the ability to rapidly assess exposure and effect of chemical contamination. Because gene expression is a sensitive endpoint, DNA-array technology is ideally suited to identify toxicity from acute and chronic exposures. DNA arrays can screen thousands of genes simultaneously eliminating the time and cost of traditional biomarker analysis. DNA arrays will aid in the determination of chemical mechanism of action and identify new environmental hazards by recognition of specific gene expression signatures. The researchers envision the use of such arrays to be extensive, with broad applications including meeting permit requirements for effluent release by municipalities, assessing new or improved treatment techniques, prioritizing watersheds for cleanup by game and fish agencies, and aid in determining water quality in water reclamation projects. This approach will provide critical information pertaining to exposure and effect of multiple pharmaceuticals, and aid in determination of deleterious effect and disease status. DNA arrays will do much to reduce time and cost of prioritizing remedial efforts, and serve as rational mechanisms to determine the "cost of residence" in contaminated waters. For this purpose, we have utilized our knowledge of aquatic toxicology and fish biology. Several aspects of fishes have driven our focus on these vertebrates as environmental assessment models. Fish directly occupy receiving waters and are excellent sentinels for aquatic and often mammalian species. They have biochemical and metabolic enzyme systems qualitatively and often quantitatively analogous to mammals. Thus, gene expression information produced from fish species can be extrapolated to other species. In addition, much environmental policy is determined based on "health" and population estimates of free ranging fishes.

1.3 Hypothesis and Objectives

The general premise of this research is that pharmaceuticals, at environmentally relevant concentrations will elicit molecular changes in exposed organisms that can be quantified and employed as a bioanalytical tool for exposure assessment. Due to diverse mechanisms of action and potential differences in cellular targets/receptors between aquatic biota and mammalian species, standard assays directed towards assessment of pharmaceutical efficacy and effect often will not be applicable for environmental monitoring. In this research a bioassay was developed that targets molecular events associated with pharmaceutical-induced gene induction and repression in exposed surrogate organisms. The long-term objective was to use pharmaceutical-induced gene expression and repression as a means to develop a DNA-array to assess exposure to and thus presence of these compounds in the aquatic environment. Additionally the bioassay approach permits evaluation of biological effects of pharmaceuticals and significantly enhances ongoing efforts to understand and address the short and long-term consequences of these compounds in the environment. To address these goals we have pursued the following specific aims:

Specific Aim I: Pharmaceutical expression profiling: Identify and clone pharmaceutical responsive genes

Task 1. Use suppressive subtractive hybridization to obtain cDNA sequences differentially expressed following exposure to prototypic pharmaceuticals.

Task 2. Use a PCR-based differential screening protocol to confirm identification of differentially expressed genes and eliminate selection of potential false positives.

Specific Aim II: Characterize differentially expressed cDNAs and gene expression profiles as markers of pharmaceutical exposure and effect by membrane hybridization.

Task 1. Establish linear range and sensitivity of membrane hybridization.

Task 2. Evaluate acute dosing on gene induction endpoints.

Task 3. Sequence and identify cDNAs by sequence homology analysis

Task 4. Determine linkage between gene induction and protein expression

Task 5. Evaluate feasibility and sensitivity of array analysis to wastewater matrices

1.4 Methods and Materials

1.4.1 Animal Stocks

Medaka embryos were collected from a brood stock colony and isolated for grow out. Fish were fed a casein-based, purified diet shown to be adequate for normal growth and reproduction, at a rate of 3% body weight three times daily (De Koven et al., 1992). At approximately two months fish were assessed for general health, and segregated according to gender. Gender was determined by analysis of secondary external sex characteristics, including development of lamellae on the anal fin of males, by using light contrast microscopy. Male fish between four to six months were subsequently used for exposures. Fish at this age were sexually mature and had fully developed organ systems. All fish were maintained at the Duke University Aquatic Research Facility containing 1600 sq. ft. of AAALAC-approved aquatic organism culture located in the Duke Forest. This facility is specifically designed for culture, maintenance and experimentation with fish. It is both temperature and light controlled with stainless steel racks supporting 100 plexiglass aquaria with 80-120 liter capacity. These house medaka of the orange red variety, and our see-through medaka brood stock. Ammonia, nitrites and nitrates, oxygen, hardness and alkalinity were routinely measured. A dedicated well provided water that was filtered over charcoal and reverse osmosis treated prior to use. Calcium carbonate was used to reconstitute water for culture. Other water purification procedures employed on line mechanical, bacterial, chemical and UV filters.

1.4.2 Exposures

Young adult, three-four months old, sexually mature, medaka were exposed to Ciprofibrate at 100, 50, 25, and 12.5 mg/kg by i.p. injection in DMSO to establish a base line LD50 value and appropriate dosing concentration and for PMP70 immunodetection studies. Based on literature values and personal communication, targeted concentrations were chosen that are sub-lethal, yet sufficient to initiate transcription of responsive genes. Each exposure group consisted of 5-10 fish and proceeded for a total of 48 hours. Environmental conditions were rigorously maintained by partial submersion of beakers into a tabletop tank system with sensitive control for light 16 hr light/8 hr dark photoperiod and temperature 25°C. Fish were fed a casein-based, purified diet shown to be adequate for normal growth and reproduction, at a rate of 3% body weight three times daily (De Koven et al., 1992). Beakers were siphoned daily to control nitrate formation and to remove deposited waste.

1.4.3 Protein Analysis

Total liver cellular protein was extracted from exposed fish by homogenization in buffer containing 50 mM Tris pH 7.6, 12 mM monothioglycerol, 1.0 mM EDTA, 1.0 mM DTT, 20% glycerol and 1.0 mM PMSF with a polytron tissue homogenizer. Subcellular fractions were made by differential centrifugation. Nuclear fractions were prepared by centrifugation of homogenates at 750 x g for 10 min, mitochondrial fraction was prepared by centrifugation of the post nuclear supernatants at 24,500 X g for 20 min, and microsomal fractions were prepared by centrifugation of the post mitochondrial supernatant at 100,000 x g for 60. Each fraction was analyzed by SDS-PAGE and immunoblotting. Protein concentration was measured by the Bradford method with bovine serum albumin as standards. Protein samples were separated by electrophoresis through denaturing 10% SDS polyacrylamide gels. After electrophoresis, protein was transferred to nitrocellulose membranes (Boehringer Mannheim, Indianapolis, IN) by electrobloting. Membranes were initially blocked in 5% (w/v) non-fat powdered milk in Tris buffered saline/Tween. Membranes were then incubated with a primary monoclonal antibody to PMP70 in 5% (w/v) non-fat powdered milk followed by washing in Tris buffered saline/Tween and incubating with alkaline phosphatase conjugated anti-rabbit secondary antibody. Protein targets were visualized with CDP* chemiluminescent substrate (Boehringer Mannheim).

1.5 Preliminary Studies and Project Preparation

Before arrays were constructed it was necessary to establish several criteria necessary to initiate fish exposures. First was the necessity to develop IACUC protocol to ensure animal health and safety guidelines. Once completed we subsequently established a broodstock of fish that could provide sufficient numbers of experimental animals throughout the duration of the research period. Secondly we conducted two pilot studies including a dose range finding study to identify maximal tolerated dose for i.p. DMSO and ciprofibrate exposures, and a pilot study examining expression of PMP70 as a biomarker of peroxisome proliferation.

1.5.1 IACUC Protocol and Developing Sufficient Fish Stock

An initial necessity for experimentation with medaka, an Animal Care and Use protocol was submitted to Duke University's Institutional Animal Care and Use Committee (IACUC), Office of Environmental Health and Safety (OEHS), and The Division of Laboratory Animal Resources (DLAR). This application was necessary to ensure the humane care, ethical treatment of all laboratory animal research, and the safety of all investigators. All protocols were approved by July 30, 2002.

1.5.2 Pilot Study: Range Finding

The researchers initially ran a range finding study to determine approximate LD50 value for anticipated antihyperlipidemic compounds used in this study including ciprofibrate, cholfibrate and chofibric acid. Preliminary results with ciprofibrate suggested that a concentration range of 25-50 mg/kg was optimal to maintain reduced mortality rates associated with i.p. exposure (Table 1-1). Initial exposures were conducted at a concentration of 100 mg/kg based on both literature values and personal communications. However when dosing medaka fish, high mortality was observed at this concentration.

Table 1-1. Range Finding, Dose Response Analysis for Ciprofibrate in Individual Medaka Fish: Survival

Compound	Ul Injected	12 Hours	24 Hours	48 Hours
DMSO	1	5/5	5/5	5/5
DMSO	2	5/5	5/5	5/5
DMOS	3	2/5	2/5	1/5
CF (12.5 mg/kg)/DMSO	1	5/5	5/5	5/5
CF (25.0 mg/kg)/DMSO	1	5/5	5/5	5/5
CF (50.0 mg/kg)/DMSO	1	7/7	6/7	6/7
CF (100.0 mg/kg)/DMSO		4/6	1/6	1/6

Interestingly Baldwin et al., (1993) suggested that medaka might not be sensitive to peroxisome proliferating agonists such as the "fibrate-antihyperlipidimeics" based on S phase synthesis activity as specific marker to mitogenic stimuli. Based solely on lethality, the researchers demonstrated that medaka respond to ciprofibrate at lower concentrations than those used in trout and medaka hepatocyte study. While this sensitivity may not be due to similar mechanism of action, these results clearly demonstrate "responsiveness" and a necessity to pursue differential toxicities in diverse organisms. Studies examining the effect of peroxisome proliferators have demonstrated significantly different actions between mammals and fishes (Cancio and Cajaraville, 2000)

1.5.3 Pilot Study: PMP-70 Study

To determine if ciprofibrate and fibrate drugs act as PPARα agonists in medaka the researchers conducted a pilot study examining the expression of PMP70, an enzyme highly induced following peroxisome proliferation. PPARα agonists have been demonstrated to significantly increase peroxisome proliferation and maybe associated with their ability to reduce plasma cholesterol levels. Peroxisome proliferation/activation however, is highly species specific suggesting that there may be multiple mechanisms of action for the fibrate-based pharmaceuticals. These studies have thus been designed to include the immunodetection of PMP-70, as a biomarker of effect for antihyperlipidemic's exposure. This extra step has enabled the assessment of biological activity for compounds tested, has served as an indicator of transcriptional alteration, and used as a benchmark to compare effectiveness of the subtractive hybridization and microarray process.

Medaka fish were exposed to either ciprofibrate or clofibrate at 12.5, 25, 50, and 100 mg/Kg for 48 hours. Following exposure, fish were anesthetized and livers removed for PMP70 analysis. Crude peroxisomal preparations were made by differential centrifugation. Protein preparations for each fraction including nuclei, peroxisomes, and post 15,000 x g spin (crude supernatant) were loaded and separated on a 10% acrylamide gel. Gels were transferred to nylon membranes and incubated with primary monoclonal antibody to rat PMP70. Analysis of gels demonstrated that exposure to both ciprofibrate and clofibrate resulted in over expression of an immunoreactive band at ~ 65 kDa in the peroxisomal fraction. Expression was highest at a dose of 25 mg/kg ciprofibrate and 25-50 mg/kg clofibrate. No immuno-reactive band was present in nuclei or post 15,000 x g supernatant.

These studies suggest that medaka may be sensitive to fibrate drugs and elicit a similar mechanistic response as that demonstrated in some mammalian species. Fibrate based drugs are thought to act by binding nuclear receptor PPAR and initiating specific gene transcription events. Whether this mechanism is functional in medaka is discussed further in chapter two. One caveat to these studies however is the size difference in the immuno-reactive band observed. Given the size difference the researchers cannot unequivocally designate this response as PMP70. Factors in their favor however are 1) antibody is monoclonal and 2) the researchers have previously observed differences in protein size when comparing mammalian and piscine homologues.

1.6 Study Plan

The overall study plan and sequence of events is described in Figure 1-1. As described above, initially it was necessary to establish and grow out sufficient animals for the beginning, middle and end stages of the research project. Given that in the facilities the researchers continuously breed multiple strains of medaka, sufficient numbers of test organisms were consistently available. Pilot studies were conducted to determine dose range response of medaka fish to fibrate compounds-where no literature values were available. These studies were necessary to identify a sub-lethal dose that would result in known molecular response that could be quantified. For this purpose they examined expression of PMP70 as a biomarker of action based on prior knowledge of the mechanism of action for the fibrate pharmaceuticals. Once a dose range had been established, medaka were exposed, tissues and RNA extracted, and processed for subtractive hybridization, differential screening, sequencing analysis and gene annotation as described in chapter two. Detailed discussion of methods, results, and conclusions associated with microarray fabrication, testing, and validation as well as identification and analysis of gene expression profiles for ciprofibrate and clofibrate are provided in Chapter 3.0.

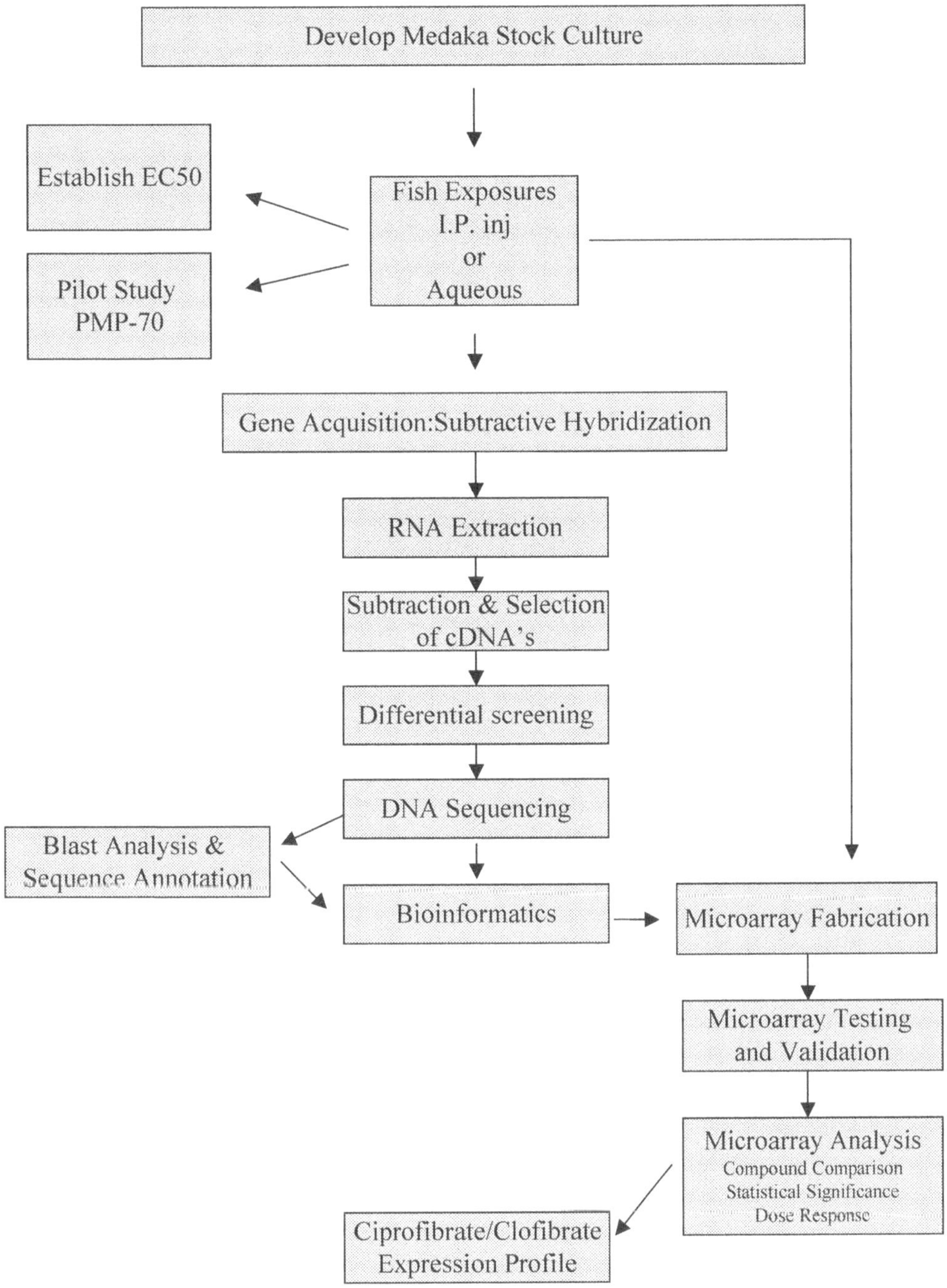

Figure 1-1. Study Plan of Events and Process.

CHAPTER 2.0

SUPPRESSIVE SUBTRACTIVE HYBRIDIZATION: ACQUISITION OF MEDAKA SPECIFIC cDNAS

2.1 Introduction

Chapter 2.0 is an in-depth analysis of the methods, results, and discussion of the subtractive hybridization procedure used to identify differentially expressed genes following exposure to ciprofibrate, the prototypic antihyperlipodemic compound belonging to the fibrate class of pharmaceuticals. The focus of this chapter is to demonstrate that exposure of medaka fish to fibrate pharmaceuticals will result in mechanism based (i.e. PPARα agonist) differential gene expression. Individual genes are subsequently sequenced, annotated and organized according to biological process. Based on frequency of occurrence and annotation category, specific mechanism of action can be configured that will facilitate identification of target-candidate genes for monitoring purposes.

Differential expression libraries were produced by suppressive subtractive hybridization (SSH) (Diatchenko et al., 1996). Subtractive hybridization enables the comparison of two populations of mRNA and the isolation of genes that are expressed in one population but not the other. SSH is a highly effective and efficient method to enrich for differentially expressed rare gene transcripts. Based on suppression PCR this technique combines normalization and subtraction in a single procedure. The normalization step is required for equalization of abundant cDNAs within the target population and the subtraction step excludes common sequences between control and treated cDNAs. Libraries of cDNAs obtained in this fashion correspond to both induced and repressed genes associated with exposure. Subtracted genes are represented as a library of PCR products which were subsequently cloned, arrayed on nylon membranes, and characterized as gene markers of exposure. By this process we have identified novel DNA sequences not previously shown to be responsive to pharmaceutical exposure in aquatic organisms. Use of ciprofibrate in this study was designed to illustrate proof of principle.

It is our hypothesis that a component of each expression library will be unique to each pharmaceutical class. These components will then be exploited as specific prognostic markers for indicating presence of and exposure to pharmaceuticals and aid in identification of molecular endpoints and biomarkers.

2.2 Materials and Methods

2.2.1 Animal Exposures

Male fish four months in age were weighed and exposed to ciprofibrate at a concentration of 50mg/kg. This dose was shown in pilot studies (see Chapter 1.0) to initiate peroxisome proliferation the intended cellular response for this class of pharmaceutical agents. Fish at this age were sexually mature and had fully developed organ systems. Exposure was by I.P. injection of ciprofibrate in DMSO using ultra thin gauge needle designed specifically to maintain health and wellbeing of fish and results in minimal tissue damage. Each exposure group consisted of 10 fish and proceeded for a total of 48 hours. Environmental conditions were rigorously maintained by partial submersion of beakers into a tabletop tank system with sensitive control of photoperiod (16hr light/8 hr dark) and temperature 25°C. Beakers were siphoned daily to control nitrate formation and deposited waste. Following exposure, whole fish or selected organs were recovered, by micro-dissection, flash frozen in liquid nitrogen and store at –80C° until processed for RNA extraction.

2.2.2 RNA Extraction

Following exposure, total RNA was isolated from whole fish or ten pooled organs (either liver, testis, or brain) by the single-step acid guanidinium thiocyanate/phenol method (Chomczynski and Sacchi, 1987). Total RNA was recovered by homogenizing tissue samples in the presence of RNA-Stat-60 (TEL-TEST, Friendswood, TX), washed with chloroform and precipitated as per manufactures recommendations. To aid in RNA quality, and remove residual/contaminating DNA, total RNA was subsequently loaded onto silica-based membrane columns (Qiagen, Valencia, CA.), treated with DNAse for 30 minutes, washed in appropriate buffers and eluted in TE buffer. RNA quality was checked on an Agelent Bio-analyzer.

2.2.3 Subtractions (SSH)

cDNA synthesis, library construction, and SSH were performed according to the manufacture's protocol for the PCR select cDNA subtraction kit (Clonetech, Palo Alto, CA.). Because of low amounts of total starting RNA, the subtraction process was performed with amplified cDNA in place of mRNA. Briefly, cDNA was prepared from total RNA by reverse transcription and amplified by long distance PCR. Amplification products were subsequently digested with RSA I restriction enzyme, purified and ligated to specific adapters in preparation for subtraction. For subtraction, cDNAs from treated fish were used as tester DNA and non-treated cDNA as driver to generate subtracted libraries that are enriched for genes that have been induced (forward subtraction). A second round of SSH was performed using cDNA from control fish as tester DNA and cDNA from treated fish as driver to generate a subtracted library enriched for genes that are repressed (reverse subtraction). Actual subtraction process was performed by a series of steps including hybridization of tester and driver cDNAs (heating DNA followed by slow re-annealing at 68°C) followed by two rounds of nested polymerase chain reaction.

cDNA subtraction libraries were cloned into a TA-pCR2.1 plasmid vector (Invitrogen), transfected into TOP10 cells and plated on agarose–AMP plates. Individual colonies were selected, grown in 96-well plates containing LB-AMP, and incubated overnight at 37°C. A total of two 96-well plates equaling 192 clones were randomly chosen for each subtraction library made (whole fish or tissue specific). For each tissue both a forward and a reverse library was constructed totaling 384 clones/tissue. Three tissues were chosen for this study including liver, brain and testis, totaling 852 isolated clones. Each clone selected and grown in this capacity represents one differentially expressed cDNA transcript. Replicate plates were made using a 96 pin-replicating tool, grown overnight and glycerol was added for storage and preservation of clones.

Differential screening of subtracted libraries was conducted using the PCR Select Differential Screening Kit (Clonetech). Selected clones from subtracted libraries were arrayed as individual colonies and hybridized to the forward (tester) subtracted cDNA library and the reverse (driver) subtracted cDNA library. Both libraries were labeled by random priming with [α-^{32}P]dCTP and unincorporated dNTPs were removed by size exclusion chromatography. Subtracted clones from each 96-well plate were PCR amplified using primers specific for adapter regions added during the subtraction process. PCR products were purified by column chromatography and transferred to two replicate nylon membranes using a hand held multi-print applicator (VP Scientific). Membranes were prepared for hybridization by denaturing, neutralizing, and fixing the DNA to the membranes by UV illumination. Hybridization of membranes with either the forward probe or the reverse probe was conducted in ExpressHyb hybridization solution (Clonetech) with 10^7 CPM per 100 ng of subtracted cDNA at 72°C overnight. Following hybridization, membranes were washed in SSC under low and high stringency. Membranes were subsequently placed into phosphorimaging cassettes (Molecular Dynamics) and exposed for varying lengths of time to achieve the best signal to noise ratios. Results of the differential screening process were based on the identification of parallel hybridizations. Replicate clones arrayed on membranes in a grid pattern enabled a direct comparison of hybridizations with forward and reverse probes. Clones that hybridize to the forward probe and not the reverse probe correspond to differentially up regulated genes. Clones that hybridized to the reverse probe but not the forward corresponded to down regulated genes. Clones that hybridize equally to both probes were considered false positives. Clones that hybridize to both probes but with different intensities were considered differentially expressed if signal intensity was 1.5 higher with either probe. Clones that did not hybridize to either probe usually represent non-differentially expressed cDNAs.

2.2.4 DNA Sequencing

All nucleic acid sequencing reactions were performed by the Center for Genome Technology (Duke University Medical Center), on plasmid template DNA using Big Dye terminator cycle sequencing chemistry (Applied Biosystems Inc). Sequencing reactions were initiated using oligonucleotide primers specific for F1 and R1 priming sites within the adaptor region of each clone, added during the subtraction process. Sequences were analyzed on an Applied Biosystems model 3700 DNA sequencer using Phred-Phrap analysis. Following sequence acquisition, all sequences were quality clipped based on a Phred-Phrap default error value of 0.05. This process ensures that all sequence information is of a high quality based on signal intensity and low background interference. Following the clipping procedure, sequences were compared to the National Center for Bioinformatics (NCBI) non-redundant gene sequence repository using either BLAST-X or BLAST-N analysis. All sequence comparisons were conducted as a batch analysis using a perl script written by Mark DeLong at the Center for Genome Technology that links homology values to the NCBI database in an HTML format. DNA sequence analysis including comparisons and subsequent alignments, were performed using SeqWeb, a front-end interface to the Wisconsin GCG suite of sequence analysis tools. SeqWeb is available to Duke University faculty and staff through the Duke Bioinformatics Shared Resource (DBSR) via a North Carolina Biotechnology Center grant. The DBSR locally maintains daily updates of the major sequence databases, and a large collection of additional public domain and commercial programs that are used to supplement the core set of tools provided by GCG.

2.3 Results and Discussion

2.3.1 RNA Extraction

The researchers have developed a process for extraction of high quality intact, non-degraded total RNA from medaka whole fish as well as specific medaka organs. This method entails a combination of single step RNA extraction using a modified and refined RNAzol agent. RNAzol promotes formation of complexes of RNA with guanidine and water molecules, abolishing the hydrophilic interactions of DNA and proteins. A second step was included to ensure minimal DNA contamination and enrich for RNA's > 200 nucleotides. Total RNA preparations from the first phase extraction were added to silica gel based membranes (Qiagen). RNA is washed in appropriated buffer loaded to columns for on-column DNAse digestion. This step ensures that there is minimal DNA contamination and selects for RNA with higher molecular weights, thus eliminating rRNA and tRNA. They have found that taking these extra steps ensures a clean RNA preparation and greatly facilitates subsequent processes including RT-PCR and cDNA subtractions. RNA quality was determined on an Agelent Bio-analyzer and is identified by the presence of 28S and 18S ribosomal RNA bands in a ratio of approximately 1.5-2:1. This is an indication that little RNA degradation has occurred and that messenger RNA is intact.

2.3.2 Subtractions: Whole Fish

The researchers original intent to perform SSH with RNA extracts from whole fish did not turn out to be fruitful. Following the SSH protocol primer sets for medaka β-Actin, were used to conduct subtraction efficiency tests. In each instance the results of this test suggested that the subtraction process was not functioning properly. They thus concluded that subtraction using whole fish RNA may not be plausible. RNA pools isolated from whole fish have been difficult to subtract and may be too complex to effectively normalize. This is most likely due to a dilution effect of induced/repressed genes from specific organs by the total mRNA pool. Theoretically, this will reduced the frequency of single stranded DNA's to find a hybridization match and will result in a loss of low abundant genes and lack of an effective subtraction. After consultation with Dr. L. Diatchenko, originator of the SSH methodology now residing at University of North Carolina Chapel Hill, she concurs with this conclusion. The alternative approach has been to perform SSH with less complex mRNA source including individual tissues.

2.3.3 SSH: Individual Tissues

Following SSH, 384 clones were randomly selected from each tissue. These clones represented potential genes that were either differentially up or down regulated by ciprofibrate exposure. The researchers have chosen to look at gene expression in liver, brain and gonad. These tissues were chosen based on known tissue targets and specific mechanisms of action of the fibrate drugs. Liver is a principle site of PPARα expression, a well-defined nuclear receptor that specifically binds fibrate pharmaceuticals and mediates differential gene expression associated with lipid metabolism and transport. They have additionally included brain and testis for assessment in the experiments due to unique function of lipid signaling in these tissues. While not major sites of PPARα expression, these tissues may represent a unique site of function of these drugs and thus potential markers of their exposure /effect. Individual colonies were selected from both forward and reverse subtraction libraries and grown in 96-well plates. These clones corresponded to 192 randomly selected induced gene transcripts and 192 randomly selected repressed gene transcripts from each tissue resulting in 384 clones selected for each tissue-a total of 852 clones for all three tissues. As illustrated in Figure 2-1, PCR products of individual clones were assessed by gel electrophoresis to determine insert size and quality of

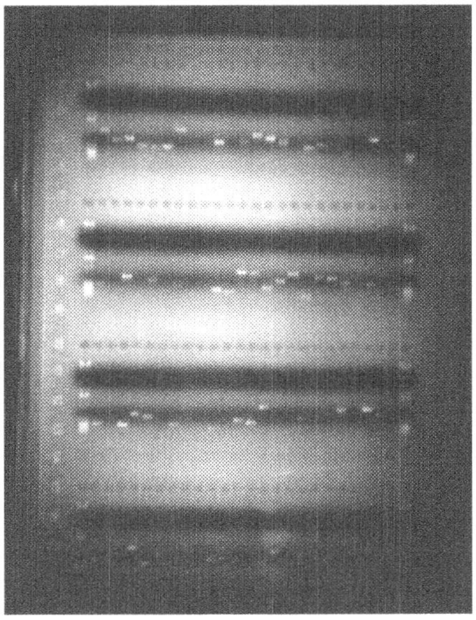

Figure 2-1. PCR Amplification of Randomly Selected Clones.
Forward SSH library medaka ciprofibrate exposed liver. PCR amplification using primers F1 and R1 to SSH adaptor sequences incorporated into each cDNA. 1.5% TBA gel, 100 volts for 1 hour.

each PCR reaction. When PCR amplified, most clones demonstrated an appropriate size and quantity for subsequent analysis (i.e. differentially screening, sequencing and array hybridization). Some clones, however, produced multiple bands following PCR amplification. This result may be due to either multiple priming sites in the cDNA sequence or selection of multiple colonies during the colony selection process. Given the abundance of colonies on any agar plate, the latter is a more likely explanation. When any of these clones proved to be of interest, they were subsequently plated for single colony isolates and separately analyzed. As false positives are common with SSH libraries, the presence of a clone in this library is not proof of its differential regulation, thus the next step was to perform a differential screening procedure. This analysis is a necessary step to ensure and independently verify differential expression of specific transcripts. Using the 96-well plate format we were able to screen 96 selected colonies at a time for false positive and false negative selection.

2.3.4 Differential Screening: Selection of Positive and Negatively Regulated Clones

Often subtracted libraries contain some cDNAs that correspond to mRNAs common to both control and treated samples. Following the production of a subtracted cDNA library, it is thus important to confirm that individual clones are truly representative of differentially expressed genes. This is typically accomplished by northern blot analysis with randomly selected subtracted clones. This process however can be tedious and time consuming. Another approach to minimizing background (i.e. false positives) is differential screening. In this method subtracted libraries were hybridized with both forward (original tester) and reverse (original driver) cDNA probes.

Clones that represent true positives will only hybridize with the forward (tester) probe. Those that hybridize with reverse probe may be considered background. Removal of false positives in this fashion provides a mechanism to enhance selection of the differentially expressed population of clones, aids in eliminating redundancy and reduce clone numbers for sampling and sequence costs. Differential screening was conducted by making replicate membrane arrays of each 96-well plate and hybridizing with either labeled forward library or labeled reverse library. As described in methods section, clones were screened by a comparison of hybridization signal between forward and reverse libraries.

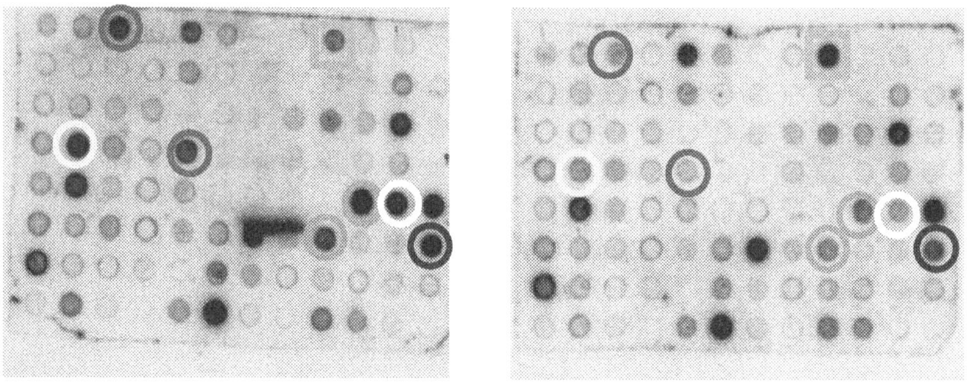

Figure 2-2. Differential Screening of SSH Libraries.
Left membrane subtracted treated library. Right membrane control subtracted library. Colored matched circles represent differentially expressed (induced) genes. Squares differentially expressed (repressed genes).

Briefly, when screening clones selected from the forward subtraction libraries, true positives will hybridize to like cDNA transcripts found in the forward library where little to no hybridization signal should be present when replicate arrays are hybridized with cDNA from the reverse library. Figure 2-2 illustrates results of the differential screening procedure. In this example, two replicate membranes representing PCR products of clones isolated from a forward subtracted liver library were hybridized with probes for either the forward subtracted library or the reverse subtracted library. Circles illustrate clones that are expressed more abundantly in the forward library than in the reverse and thus represent true positives, i.e. gene transcripts that are up regulated. The boxed clone gives the opposite to the expected result and would not be scored as a truly up-regulated gene. However, given that the hybridization signal is significantly stronger in the reverse probe this may represent a gene transcript that is down regulated (even though it is the opposite of the intended response) and the researchers have chosen to examine this (and similar clones) further in array studies. Those clones where the hybridization signal is equal between probes or no signal is present beyond background, most likely represent false positives. The differential screening procedure was conducted on all 852 randomly selected clones. Those that demonstrated a significant difference in expression were further selected. Master 96-well plates of these clone sets were made, used for sequence analysis and were used for all array hybridization studies.

2.3.5 Clone Sequencing and Analysis

Clones were sequenced as described in material and methods section. Briefly, selected clones were PCR amplified and treated with exonuclease and alkaline phosphates to digest PCR primers and phosphorylation of remaining dNTPs respectively. This treatment is recommended for clean up of PCR products prior to sequencing analysis and is a quick and inexpensive procedure that can be substituted for column chromatography. Sequencing was performed by standard big dye terminator chemistry and all clones were sequenced in both directions. Sequence information was quality clipped using the perl script written at the Center for Genome Technology Duke University. Quality clipping of sequence data is a necessary step to ensure high quality sequence information and minimize background signal., This becomes extremely important when conducting BLAST homology searches, as bad quality sequence data will significantly askew homology matching (i.e. estimated matrix scoring and "e" values) during comparisons in the database. Of the 852 original clones randomly selected during SSH, 281 clones were further chosen following the differential screening procedure for sequence analysis. Table 2-1 and Figure 2-3 illustrate initial sequencing and BLAST homology results for each tissue. Table 2-1 summarizes number of clones evaluated, and gene homologies i.e. number of sequences having significant homology to known genes in the non-redundant NCBI protein database. All sequence comparisons were made using Blast X, which compares the translated (in all six directions) protein sequence of the unknown SSH clones to all know protein sequences. The researchers have consistently obtained higher homology scores when comparing protein sequence than with DNA sequence (BLAST-N). This is most likely due to differences in codon bias between species. In liver, a total of 124 clones were sequenced. Of these 124, 87 clones had significant homologies to existing "named" proteins.

Homologies are based on a comparative matrix value to all known and available sequences. The lower the "e" values, the higher the homology-match. A value of "0" is a perfect match. We have defined a cut-off value of $1e^{-10}$ for assignment of significant homology. Values below $1e^{-10}$ demonstrate very little homology and thus are not representative of true orthologus sequences. 18 genes isolated from liver did not show any significant homologies to the protein database. In an effort to determine if we had sequences representing the 5' or 3' un-translated

Table 2-1. Suppressive Subtractive Hybridization Summary for Ciprofibrate-Injected Male Medaka Organs.

	Brain	Liver	Testis
Total colonies screened	384	384	384
True positive colonies sequenced	66 (17.2%)	124 (32.3%)	88 (22.9%)
Poor sequence data	19 (4.9%)	27 (7.0%)	24 (6.3%)
Positive sequence data	47 (12.2%)	97 (25.3%)	64 (16.7%)
No significant homology	17 (4.4%)	10 (2.6%)	18 (4.7%)
Significant homology, redundant	30 (7.8%)	87 (22.7%)	46 (12.0%)
Significant homology, non-redundant	29 (7.6%)	35 (9.1%)	42 (10.9%)

All percentage values are relative to the total colonies screened (384) from each organ region of know genes , the researchers re-ran BLAST-N comparisons for each unknown sequences to the nucleotide database. This procedure usually resulted in significant homologies for one or two of the unknown sequences. The remaining sequences thus represent unique gene sequences not previously identified or annotated in any other species (See Figure 2-3). An additional number of clones also resulted in failed homology matches. Further analysis of raw sequences data suggested that most of these clones contained a high background and thus no consistent quality sequence could be delineated.

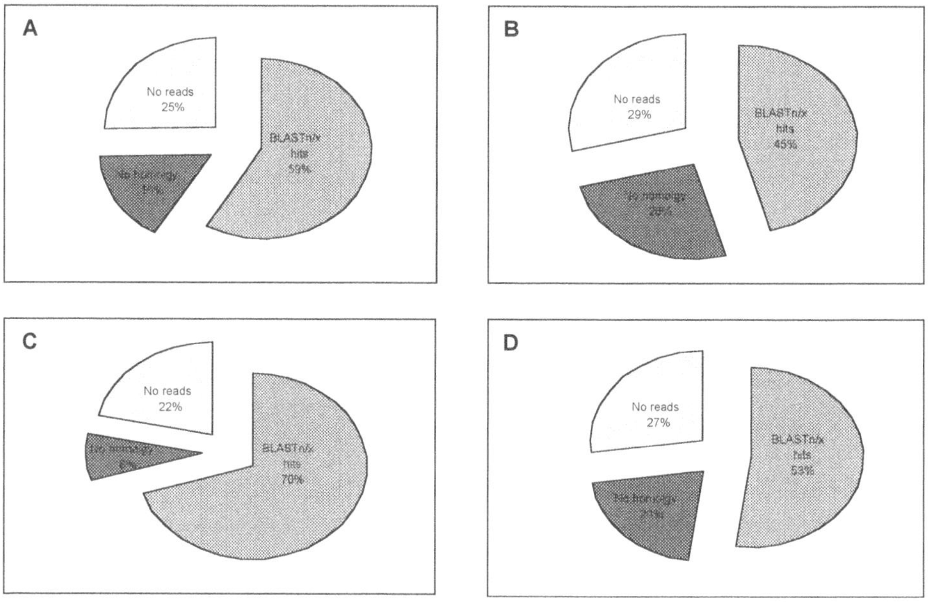

Figure 2-3. Pie Chart Illustrating Distribution of the A) Total, B) Brain, C) Liver, and D) Testis Clones Sequenced.

2.3.6 Tissue Specificity

Subtraction libraries demonstrate that differential gene expression is highly tissue specific. Few genes were expressed across multiple tissues with one gene (MHC class I) common to all tissues examined. Not surprisingly, this gene is associated with immune response and is a component of the acute phase response, which, as described in more detail below may represent a more generalized stress response to fibrate exposures. Five genes were common to both brain and testis including *G(3)DPH, MHC class 1, Mitochondrial DNA, Protease (macropain) 26S ATPase, Sodium/postassium ATPase alpha1.* Annotation of these genes suggest that overlap in gene expression is not linked to a particular function or biochemical pathway and may be due to random selection of clones. These results demonstrate that ciprofibrate exposure elicits gene expression in all three tissues examined and that expression is highly tissue specific. Whether gene expression is a direct result of PPARα activation cannot be determined from this data. And while PPARα is a primary target receptor for fibrate pharmaceuticals in each of these tissues, alternate mechanisms of fibrate induced gene induction may be associated with observed results.

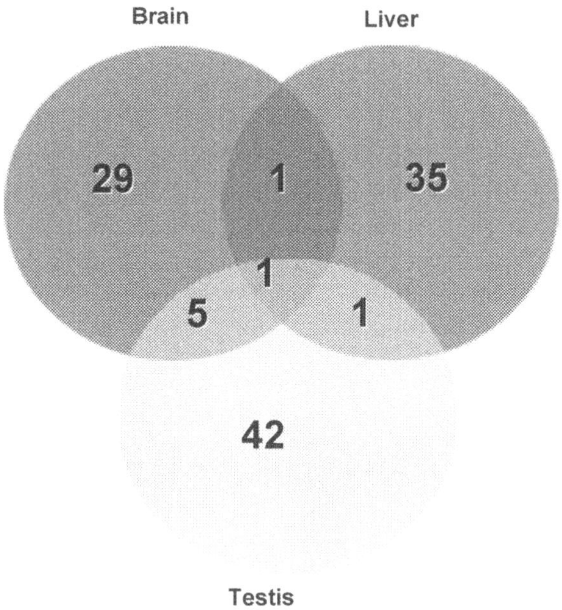

Figure 2-4. Venn Diagram Illustrating Numerical Distribution of 100 Total Non-redundent Genes in Brain, Liver, and Testis of 50 mg/kg Ciprofibrate Injected Adult Medaka.

2.3.7 Sequence Annotation

To facilitate interpretation of results, genes were grouped according to their cellular processes in which they are likely to be involved based on information from Pub Med, Swiss-Prot, LocusLink, and GO annotation databases. Genes were annotated with processes that are consistent with the pharmacological effects of fibrate drugs and PPARα agonists in mammalian studies. Most genes annotated as highly similar or moderately similar to named genes could be grouped in this fashion. Even though many genes may be involved in more than one process, each gene was grouped into only one category. Attempts were made to annotate a gene with the cellular process where it was considered most likely to belong in the context of pharmacological effects of PPARα agonists. Additionally, given that clones from SSH were picked randomly, the frequency with which particular cDNAs are found reflects the abundance on the corresponding mRNA in the original tissue extract. Gene names and frequencies are provided in the Appendices.

Results obtained from the sequencing analysis suggests that ciprofibrate exposure elicits interesting molecular changes and pathway relationships. Annotation of gene expression suggests that in all three tissues ciprofibrate regulates the expression of a number of genes involved in the metabolism and transport of lipids, glucose homeostasis, amino acid metabolism, cell proliferation and differentiation, and apoptosis. Additionally and interestingly, gene response in liver also encompassed specific components of the immune and inflammation response.

2.3.7.1 Fatty Acid Metabolism

Ciprofibrate and other fibrate drugs belong to the antihyperlipidimics (hypolipidemic) drug class. Their main prescribed function is for amelioration of metabolic disorders including hyperlipidemia and cardiovascular disease. These disorders have a complex etiology involving both genetic and nutritional factors. In mammalian species, the molecular mode of action for the fibrate drugs are as ligand specific agonists to the nuclear receptor PPARα. Molecular changes associated with PPARα agonists have been well described (Yadetie et al., 2003;Yamazaki et al., 2001), and with recent microarray data the influence of PPARα agonists on liver system function is coming to light. The list of known PPARα targets encompasses an array of genes that participate in numerous aspects of lipid metabolism and transport, including fatty acid uptake in membranes, fatty acid binding in cells, fatty acid oxidation (peroxisomal and mitochondrial) and lipoprotein assembly and transport (Kersten et al., 2001). PPARα is a member of the nuclear receptor family of transcription factors, a diverse group of proteins that mediate ligand-dependent transcriptional activation and repression. Several studies have shown a direct involvement of PPARα in the fibrate-modulated gene expression of hepatic apo A-I and apo A-II, (the major apolipoproteins in high-density lipoprotein (HDL)), of lipoprotein lipase and apo C-III, (both major determinants of plasma triglyceride levels), and of several enzymes implicated in fatty acid β-oxidation such as acyl-CoA oxidase (ACO).

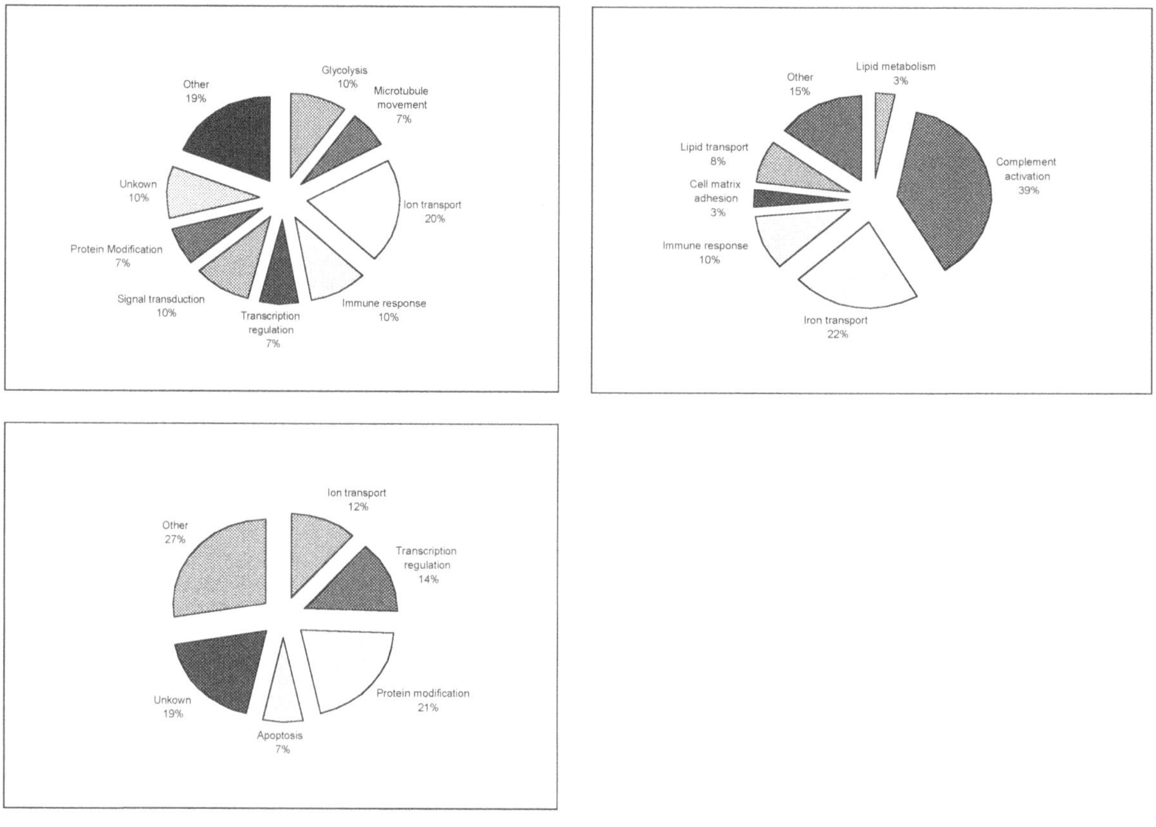

Figure 2-5. Gene Ontology™ (GO) Distribution of Differentially Expressed Genes in 50 mg/kg Ciprofibrate Injected Adult Male Medaka in A) Brain, B) Liver, and C) Testis. Go Terms Based on General Biological Processes as Determined by LocusLink (NCBI).

In this study, several genes associated with lipid metabolism and transport, were identified as differentially expressed following ciprofibrate exposure. In liver, apolipoprotein A and B and basic fatty acid binding protein were identified indicating an influence of ciprofibrate on lipid transport in medaka liver. PMP70 was shown to be up regulated in testis. Identification of PMP70 differential gene expression suggests that the molecular mode of action for ciprofibrate i.e. PPAR agonist, is functional in medaka. To date however, whether fish share an exact pharmacological response as mammals is to be determined. Regardless, the high degree of PMP70 up regulation suggests that peroxisome proliferation is an immediate effect of ciprofibrate exposure and can be exploited as one of the intended biomarkers for this drug class.

2.3.7.2 Immune and Inflammation response-Acute Phase Proteins

One of the more interesting responses observed following ciprofibrate has been the identification of genes associated with immune function and inflammation (See Appendices A1-A3). The acute phase response (APR) is a systemic inflammatory component of innate immunity. APR is initiated by chemical patterns that are presented by invading microorganisms or revealed by tissue damage resulting in the production of pro-inflammatory cytokines. These cytokines in turn modulate immunity by inducing changes in gene expression of key metabolic pathways. Systemic inflammation is accompanied by an increase in triglyceride-rich lipoproteins, a reduction in high-density lipoprotein (HDL) cholesterol and impairment of cholesterol transport (Bayne and Gerwick, 2001). Overall the APR results in metabolic adjustments and suppression of cholesterol, fatty acid and phospholipids biosynthesis. Activation of PPARα by fibrates is thought to have an anti-inflammatory effect. The mechanism of this action is proposed as a inhibition of inflammatory cytokines from macrophages and monocytes by antagonizing the activities of the transcription factor, nuclear factor-kB (NF-kB) (Pointer and Daynes, 1998). Microarray data has demonstrated the coordinated down regulation of numerous genes associated with the APR following ciprofibrate treatment (Yadetie et al., 2003).

In these studies, numerous genes representing an acute phase response were observed specifically pentraxin/C-reactive protein, components of the compliment system and the major histocompatability complex. Additionally, as demonstrated in the section above, genes associated with fatty acid metabolism and transport, are integral components of the acute phase response. APR proteins can be categorized into four distinctive groups. Group 1: Minor or Intermediate-protein concentrations increase one to two fold, Group 2: Intermediate-protein concentrations increase three to five fold, Group 3: Major-protein concentrations increase up to 1000 fold. Group 4: Negative-protein concentrations decrease. Of those genes down regulated by ciprofibrate treatment in medaka, pentraxin/C-reactive protein was extremely responsive and highly down regulated. C-reactive protein is considered to be a major Group 3 APR protein with levels increasing up to 1000 fold (Bayne et al., 2001). No consistent regulation was observed for complement 3 (Group 2), however complement 9 and C1r were both observed to be up regulated. Complement H factor gave a strong and specific induction response. Compliment H is an inhibitory factor that inactivates the alternative complement system preventing damage to normal host cells. C1q-like adipose specific protein-also called percerebellin like protein was down regulated. Percerebellin has previously been found as an up regulated protein during acute phase response in trout (Gerwick et al., 2000). Structurally, percerebellin resembles the collagen portion of the C1q complex of A, B, and C chain compliments. In mammals, percerebellin is expressed only in brain, thus the relevance of this gene product to acute phase response in fish liver is unclear. Overall the immune and inflammation gene response to ciprofibrate in medaka liver is consistent with the anti-inflammatory activity of fibrate drugs in mammals.

2.3.7.3 Sugar Metabolism

Ciprofibrate exposure resulted in differential regulation of genes encoding enzymes involved in sugar metabolism, including Aldolase A, Aldolase B, G3PDH, PEP carboxylase and Enolase. Each of these enzymes is associated with an important step in glycolysis. Microarray studies of peroxisome proliferating agents show a consistent modulation of glycolytic enzymes

(Yadetie et al., 2003). Fasting PPARα knockout mice develop severe hypoglycerima, suggesting a role for PPARα in glucose homeostasis (Kersten et al., 1999). This effect is thought to be secondary to reduced fatty acid oxidation, which effects hepatic gluconeogenesis (Leone et al., 1999). Given the interrelationship between fatty acid, lipid and glucose metabolism and associated metabolic diseases such as diabetes, modification of glycolytic enzymes by ciprofibrate is of no surprise.

2.3.7.4 Cell Processes

Many enzymes with differing function were lumped into the cell process category. It is unknown at this point and time if genes regulated in these categories are direct effects of ciprofibrate activity as a PPARα agonist, an alternative mechanism of action or are indirect effects due to modification of other metabolic pathways. It is important in any differential gene expression study to discern immediate (direct) effects to those that may be secondary (indirect). Given the known mechanism of action for this class of drugs, we can speculate to some degree on direct actions, including those associated with lipid metabolism and transport. In the cell processes category we have identified genes associated with cell attachment, mitochondrial function, apoptosis, signal transduction, sex determination, and others. Even when grouped according to tissues type, no clear pattern of function or process emerges. Several genes in this category however demonstrate strong regulatory effects following ciprofibrate exposure. In the brain, muscle type creatine kinase (CKM1) is highly up regulated. CKM1 transfers a high-energy phosphate group to creatine. Phosphocreatine acts as a reserve source of high-energy phosphate groups. In testis, Cathepsin C is also highly up regulated. Cathepsin C is major proteolytic protein found in lysosomes and is responsible for degradation of the basement membrane.

2.3.7.5 Metabolism, Structural, Protein Degradation, Growth and Proliferation Transcription and Translation and Unknown

Several genes from the remaining categories are worthy of noting here. In liver, transferrin and serotransferrin were expressed in high abundance. The frequency of expression reflects a common differential gene expression observed with several compounds in our laboratory, including estrogen and environmental estrogens. Transferrin belongs to a family of proteins associated with iron transport; it also is significantly down regulated as a component of the acute phase response. Given that ciprofibrate exposure initiated differential regulation of other components of the acute phase response, we speculate that modulation of these iron transport proteins is coupled to the APR. Metabolism genes for both phase one and phase two were additionally isolated. Interestingly, CYP3A was found but not CYP4. CYP4 in mammals and some fish species is associated with ω-fatty acid hydroxylation and is up regulated by peroxisome proliferators. As a major emphasis of study in our laboratory, we have demonstrated that CYP regulation in fish often deviates from that observed in mammalian species. The lack of CYP4 response will be worthy of studying in the future.

Alpha tubulin was significantly up regulated in both the brain and the testis. Alpha tubulin is a structural gene associated with the cytoskeleton–microtubule network of the cell and is important for intracellular trafficking of chromosomes, golgi bodies and proteins. Long thought as a constitutively expressed "housekeeping" gene, these studies and others conducted in our laboratory show that this gene is specifically down regulated. Often alpha tubulin is used to normalize gene expression data, however, these studies demonstrates that use of tubulin expression in this capacity would result in a flawed data set.

Of the transcription factors identified both the retinoid X receptor (RXR) and SP1 were shown to be down and up regulated respectively. RXR belongs to the nuclear receptor gene family and acts as a dimerization partner to other members of the NR family. Specifically, RXR heterodimerizes with PPARα following ligand binding. The receptor complex subsequently binds with its cognate DNA response element to facilitate gene transcription.

2.4 Concluding Remarks

In this component of the study the researchers have completed a multi-organ toxicogenomic analysis of the antihyperlipodemic compound ciprofibrate in the medaka fish. The objectives were two fold; 1) assess the utility of medaka fish as a comparative, genomic, vertebrate model for expression profiling of candidate pharmaceutical ligands, and 2) acquire genes that could be used for the development of an aquatic gene array. Subsequently this array would be employed as a bio-analytical tool for identification of pharmaceuticals in reuse waters and aquatic environments. Given that no commercial arrays are available for medaka, the researchers first conducted a tissue specific expression sequence tag EST project to generate medaka cDNAs. ESTs (~1000) were isolated using suppressive subtractive hybridization following pharmaceutical exposures. This approach is a sensitive approach for analysis of mRNA expression profiles by generation of biased cDNA libraries where composition and relative abundance of individual transcripts can be determined. Numerous genes have been identified as responsive to ciprofibrate exposure in each tissues/organs examined in this study. Gene expression was highly tissue specific with few common overlapping genes. In the context of developing an expression profile for ciprofibrate, and other members of the fibrate class of pharmaceuticals, distinct gene expression patterns are used to produce a defined finger print of exposure to and thus presence of these compounds in contaminated waters. Researchers discusse the identification and regulation of a few suspected key genes, gene annotation and predicted responses. This information greatly facilitates a more comprehensive understanding of the mechanisms of action following pharmaceutical exposure in medaka fish. It is likely in most cases that these organisms will share some similar gene target receptors as those identified in mammalian species. What are unknown however, are non-target mechanisms in mammals and other vertebrates that may be associated with the pharmacological action and potential aberrations following exposure. Generation of an inducible/repression gene profile will shed some light on this lack of data and aid in developing a gene set predictive for exposure assessments.

CHAPTER 3.0

MICROARRAY CONSTRUCTION AND ANALYSIS

3.1 Introduction

The impact of genomic technology on the progress of biomedical and environmental science is difficult to overemphasize. The potential rewards of the multiple species genome projects in providing detailed blueprints of genomes are enormous. Already, this information is accelerating the pace of gene discovery. Equally important has been the development of technologies to measure gene expression on a global scale. The development of DNA microarray methodology that enables the simultaneous analysis of tens of thousands of genes has transformed research that aims to understand gene regulation and pathways of gene control. This technology provides the means to assess global changes in gene expression in response to environmental toxicants. Further, changes can be explored without the need for prior knowledge or bias in the targets that are measured. Thus, it becomes a discovery tool that can identify new genes and pathways that are linked to regulatory process that may be affected by environmental toxicants. The ability to globally measure gene expression represents a unique tool for defining phenotypic characteristics in a cell or organism. The pattern of genes expressed following exposure to environmental toxicants becomes a signature of that toxicant (i.e., a biomarker of toxicant exposure). Further, the delineation of phenotypes associated with acute or chronic exposure to environmental toxicants is critical to the understanding of diverse phenomena, including the relation among genotype, environmental exposure and time in the etiology of disease. Toxicogenomics, a sub-discipline derived from the combination of genomics and traditional toxicology, offers the opportunity to develop signature biomarkers of toxicant exposure, aid in identification of common mechanisms of toxicity and to define cellular pathways of response injury and toxicant-induced diseases.

In Chapter 3.0 the researchers are focused on the development of microarrays for analysis of fibrate pharmaceuticals using the medaka fish. The general approach taken is to apply the gene sets acquired as described in chapter two, to membrane based arrays and determine and characterize differentially expressed cDNAs and gene expression profiles as markers of pharmaceutical exposure and effect. The two major tasks associated with this aim seek to establish linear range and sensitivity of membrane hybridization and evaluate acute dosing on gene induction endpoints. The ultimate goal however is to develop a set of gene markers that can be applied to identify the presence of fibrate based pharmaceuticals in reuse waters. The compounds analyzed in this study include ciprofibrate, clofibrate and chlofibric acid. The researchers have chosen these compounds as prototypic fibrate based drugs based on their mechanisms of action and reported presence in contaminated surface waters. In the first section of this chapter we report on array fabrication and dynamics including spotting of DNA targets on membranes, non-amplified versus amplified probe production, software for image analyses, and development of internal standards for normalization. Each of these factors is important and has contributed significantly to the development of a microarray for assessment of fibrate pharmaceutical exposures in medaka. In the second section of this chapter we report on a multi-factorial experimental design that incorporates dose, time and tissue and compound specific effects. In this set of experiments the researchers have expanded their array development to incorporate a larger set of medaka specific genes and have used specific bioinformatics software to aid in computational analysis of gene expression profiles. Statistical analysis was performed on all data sets to identify significance of gene responses. Below is a detailed description of this research activity and methodology developed.

3.2 Methods

3.2.1 Chemicals

Ciprofibrate, clofibrate, and clofibric acid stocks were prepared in HPLC-grade DMSO and stored at -20°C until exposures.

3.2.2 Test Animal

Oryzias latipes *(medaka)* is a small (3-4 cm) egg-laying freshwater fish native to Japan, Korea, and eastern China, and is a well-established toxicological animal model. Fish used for this study were collected from a stable orange-red line under standard flow-through aquaculture conditions. All fish were handled and treated according to guidelines set forth by the Duke Institutional Animal Care and Use Committee (IACUC).

3.2.3 Aqueous Exposures

Adult male medaka (six- to seven months-old) were isolated to a separate aquarium tank and acclimated under flow-through culture conditions for one week prior to aqueous exposure. Only males were used in this study to eliminate effects of circulating estrogens in female fish. Based on preliminary data, treatment groups (all nominal concentrations) were chosen for designed experiments: control, vehicle (DMSO) control, 0.01 mg/L ciprofibrate, 0.10 mg/L ciprofibrate, 1.0 mg/L ciprofibrate, 0.01 mg/L clofibrate, 0.10 mg/L clofibrate, 1.0 mg/L clofibrate and 0.1 mg/L clofibratic acid. DMSO and stock chemicals were spiked directly into ethanol-rinsed 2-l glass beakers containing 2 L sterile, well-aerated embryo rearing medium (ERM) (17.1 mM NaCl, 272 μM $CaCl_2 \cdot 2H_2O$, 402 μM KCl, 661 μM $MgSO_4 \cdot 7H_2O$; pH 7.2), and allowed to equilibrate for at least 1 hr. Final DMSO concentrations in vehicle and treatment solutions were ~0.01%. Beakers were covered with fine mesh and fish were incubated at room temperature under 16:8 light:dark conditions for 7 d. Fish were fed twice daily with three drops of freshly hatched artemia during the entire exposure period. Treatment solutions were 50%-renewed daily with 1 L ERM containing the appropriate concentration of chemical. No mortalities or abnormal behavior were observed during 7 days of static-renewal exposure. On days 1, 3, and 7, four fish per treatment were removed and anesthetized in ice-cold ERM. Using RNase-free scissors and forceps, organs (brain, liver, and testis) were removed and immediately frozen in liquid nitrogen. For each sampling day, four individual organs (brain, liver, or testis) from each treatment were equally divided and pooled for total RNA extraction and cDNA microarray analysis (2 organ pools per treatment per day equaling two biological replicates).

3.2.4 RNA Isolation

For each treatment specific RNA extraction, pooled organs from brain, liver, or testis, (two organs/biological replicate–two replicates) were homogenized with 1 ml RNA Bee (TelTest) using an RNase-free stainless steel Polytron homogenizer (Kinematica). Following homogenization, 100 μl chloroform was added and the homogenate mixed vigorously for 15 s. The mixture was then chilled on ice for 5 min and centrifuged at 12000g at 4°C for 15 min. The upper aqueous phase was removed and 100 μl isopropanol was added to the lower phenol:chloroform phase. Samples were incubated at 4ºC for 15 min and then centrifuged at 12000g at 4ºC for 15 min. The supernatant was discarded and the RNA pellet was washed once by vortexing with 800 µl 75% ethanol; the sample was then centrifuged at 7500g at 4ºC for 8 min. Following removal of the ethanol wash, the sample was air-dried for 10-15 min. The RNA pellet was then re-suspended in 100 µl DEPC-treated water. To eliminate DNA contamination, each sample was on-column-digested with DNase using an RNase-free DNase Set according to manufacturer's instructions (Qiagen), and eluted with 30 µl warmed (52ºC) sterile water. RNA quantity and quality was verified using a NanoDrop® ND-1000 spectrophotometer and Agilent BioAnalyzer, respectively. Total RNA concentrations ranged from 400-2000 ng/µl with a 260/280 ratio ≥1.8.

3.2.5 cDNA Microarray Fabrication and Hybridizations

Medaka brain, liver, or testis, specific cDNA microarrays were constructed with genes from subtracted (SSH) libraries generated from ciprofibrate, 17β-estradiol, or 2,3,7,8-tetrachlorodibenzo-p-dioxin treated fish. Brain-, liver-, and testis-specific cDNA microarrays contained 186, 186, and 282 non-redundant medaka genes respectively. To custom build these arrays, colonies containing these cDNA clones were individually transferred in duplicate with sterile toothpicks to a 96-well microplate containing 100-µl LB-amp media per well and re-grown overnight at 37ºC while shaking. Following overnight growth, all isolated stock colonies were stored in 20% glycerol at -80ºC until PCR amplification. All cDNA clones were simultaneously PCR-amplified in duplicate using a 96-well PCR plate. For each 75-µl PCR reaction, cDNA clones were PCR-amplified using 3 µl bacterial cells, 7.5 µl 10X Advantage 2 buffer, 2.25 µl 10 µM NP-1 primer, 2.25 µl 10 µM NP-2R primer, 2.0 µl 10 µM dNTPs, 57.25 µl sterile water, and 0.75 µl 50X Advantage 2 Polymerase Mix (BD Biosciences). PCR reaction conditions were: 94ºC for 30 s followed by 35 cycles of 95ºC for 10 s and 68ºC for 3 min. PCR-amplified cDNAs were purified in 96-well format using an ArrayIt PCR Purification Kit (ArrayIt Microarray Technology), quantified in 1 µl-aliquots using a NanoDrop® ND-1000 spectrophotometer, and diluted to 20 ng/µl with sterile water. For each cDNA membrane array, 50 ng of purified cDNA was denatured at 65ºC for 5 min with 0.6 M NaOH and chilled on ice for 2-3 min; 25 ng were duplicate-spotted using a 96-slot pin Multi-Blot™ Replicator (VP Scientific, Inc.) with a 2-µl slot pin volume onto positively charged nylon membranes (75 mm X 115 mm) (Roche Diagnostics). Spotted membranes were neutralized in 0.5 M Tris/1.5 M NaCl for 4 min, rinsed in water for 2 min, and cDNAs UV-cross-linked prior to drying and storage. To assess intra-array variability, array controls were randomly spotted (20 ng/spot) on each membrane array. Array controls included: Arabidopsis Cab1 cDNA (6 spots/array), salmon sperm DNA (4 spots/array), and human β-actin (2 spots/array) (Spot Report-3 Array Validation System; Stratagene). Two arrays per treatment for day 7 were printed for brain gene expression analysis from DMSO-, ciprofibrate- and clofibrate-exposed male medaka, yielding a total of 16 identical brain-specific membrane arrays; two arrays per treatment for days 1, 3, and 7 were printed for liver gene expression analysis from DMSO-, ciprofibrate- and clofibrate-exposed male medaka, yielding a total of 42 identical liver-specific membrane arrays; and two arrays per treatment for day 7 were printed for testis gene expression analysis from DMSO-, ciprofibrate- and clofibrate-exposed male medaka, yielding a total of 16 identical testis-specific membrane arrays. All arrays were stored in plastic bags at room temperature until hybridizations.

^{32}P-labeled cDNA probes were generated in a single labeling reaction using total sample RNA (DMSO-, ciprofibrate-, or clofibrate-treated brain, liver, or testis RNA) spiked with internal standard mRNA (Spot Report-3 Array Validation System; Stratagene). All labeling was performed with using Strip-EZ RT Kit protocol (Ambion) allowing the ability to remove hybridized RNA and reuse individual membranes for subsequent experiments. *Arabidopsis Cab1* mRNA was used as an internal standard to control for labeling and hybridization efficiency. Briefly, 2 µg total RNA, 0.5 ng *Arabidopsis Cab1* mRNA, and 2 µl 10 µM oligo(dT) primer were mixed and denatured at 65°C for 5 min. After allowing the mixture to cool to room temperature for 5 min, 2 µl 10X RT buffer (Ambion), 2 µl 10X dNTPs, 20 µCi [^{32}P]dATP (Amersham Biosciences), and 1 µl MMLV RT (Ambion) were mixed thoroughly with the above mixture, and the final mixture incubated at 42°C for 1.5-2 h. Following incubation, all ^{32}P-labeled cDNAs were purified using Microspin G-25 columns (Amersham Biosciences), and probe radioactivity quantified using a Beckman Coulter LS 6500 scintillation counter (Beckman Coulter). For double labeling reactions, radioactivity of purified ^{32}P-labeled cDNAs ranged from $0.30X10^5$-$2.5X10^5$ cpm/µl; based on many labeling reactions performed in this laboratory, we have observed that yields are typically two-fold higher when labeling reaction time is extended to two hours.

Using glass hybridization tubes, membrane arrays were individually pre-hybridized with 5 ml hybridization buffer (ExpressHyb Hybridization Solution, BD Biosciences) and 100 µl blocking solution (10X SSC/5.5 mg/ml salmon sperm DNA) at 72°C while rotating for 1 h. Equal quantities of denatured ^{32}P-labeled cDNA probe diluted in 100 µl blocking solution were hybridized to appropriate membrane arrays overnight while rotating at 72°C. Membrane arrays were washed four times with 2X SSC/0.5% SDS for 15 min/wash, and two times with 0.2X SSC/0.5% SDS for 15 min/wash. Following washes, hybridized membranes were plastic wrapped, exposed onto 35X43 cm phosphor screens for 72 h, and scanned using a Storm 860 Phosphoimager (Molecular Dynamics).

3.3 Results and Discussion

3.3.1 Microarray Array Design, Construction, Testing and Validation

3.3.1.1 Development of a 48 Gene Microarray

To initiate array studies a small 48-gene array was produced to examine specific parameters associated with membrane based microarray construction, testing, and validation. Specifically it was necessary to determine experimental parameters including array design, spotting method, gene density, spot size, spot concentration, probe labeling and preparation, feature extraction, spot segmentation, background subtraction and data normalization. Each of these variables impact microarray performance including array reproducibility and variance and the ability to produce meaningful, statistically significant data sets.

The 48-gene array was comprised of select set of genes demonstrating a strong and specific response during the differential screen selection process (see Chapter 2.0). Select clones were chosen from liver, testis and brain to broaden scope of response from each tissue. Additionally the 48-gene array included genes from differing structural and/or functional categories including fatty acid binding and metabolism, cellular processes and cell structure.

3.3.1.2 cDNA Spotting Concentrations

Studies were conducted to identify a spotting concentration sufficient enough to detect the linear range of mRNA concentrations without saturation and display a high degree of sensitivity. Duplicated membranes were constructed that contained multiple replicates of each gene with differing probe spot concentrations between 1 and 100 ng. Membrane were exposed to a constant source of control liver mRNA and spot intensity values were determined. Intensity values increased linearly with probe concentration to approximately 25ng at which no further increase in signal was detected. 25 ng per spot was subsequently chosen for all further membrane fabrication.

3.3.1.3 Examination of Probe Labeling

Using the 48-gene array, a series of experiments were conducted to evaluate differences in probe preparation. To best optimize this system, several parameters were considered 1) cDNA synthesis-via random priming or poly dT and 2) the use of amplified or non-amplified probe.

For cDNA synthesis all experiments were conducted with cDNA probe made from poly dT priming. While creating a reverse transcription bias for 3' ends of cDNA, specific activities of poly T priming were consistently higher than those produced by random priming. Additionally all gene targets (gene probes on arrays) were produced with SSH and a poly dT reverse transcription, which has an inherent 3' gene bias. Production of cDNA probe using poly dT retains this bias and minimizes any discrepancies that may occur with the random priming.

Given that all genes have been isolated from SSH, a method specific for identification of low expressed transcripts, it was necessary to address the question of probe amplification, i.e. would labeling non-amplified RNA (RNA reverse transcribed to cDNA) be sensitive enough (contain high enough message) to hybridize to the membrane or does the entire level of probe need to be amplified? Two different techniques were examined for probe production, Both techniques generated ^{32}P-labeled cDNA probes from RNA isolated from pooled tissue (i.e. livers from five identically treated male medaka). Probes were generated using either the non-amplified the Strip-EZ kit (Ambion) or the Atlas SMART Probe Amplification Kit (Clontech).

The Strip-EZ kit technique involved generating ^{32}P[dATP]-labeled cDNAs from purified total RNA using oligo(dT)s as primers with an MMLV reverse transcriptase. One benefit of this particular method was the modified dCTP included in the kit allowing for an enzymatic cleavage of the residue during the stripping procedure. This provides a more complete removal of the probe from the membrane under less intense or potentially damaging (loss of some of the spotted cDNA) conditions and enables the same membrane to be reprobed up to ten to twelve times with little to no loss of signal intensity. A concern with the above labeling technique is potentially missing real differences in gene regulation due to low signal intensities. As mentioned previously, ^{32}P-labeled probes were used to help combat this problem. An alternative technique is to use a method that will increase the signal intensity and thus the sensitivity of the array. The Atlas SMART Probe Amplification Kit (Clontech) was used to amplify probe prior to labeling and obtain higher signal intensity. The kits used to obtain the original clones during the SSH process were also from Clontech and therefore many of the components and techniques used for the probe amplification are similar and known to work for the researchers.

The results of probing the microarray with non-amplified ^{32}P-labeled probe were successful and make this technique a viable choice for future experiments. The researchers generated and used amplified probes on two separate occasions with positive results (data not shown). The SMART Probe Amplification procedure allowed us to generate probes with more activity and obtain higher signal intensities, while maintaining much (90%) of the original complexity and relative abundance of the original RNA samples, compared to non-amplified probes. While this obtained higher signal intensities with amplified probes, the sensitivity did not appear to increase. This, in part, may be due to the fact that most to all of the cDNAs spotted on the membranes were from rather strongly responding up or down regulated genes. Using amplified probes may become more important when examining low-abundance transcripts or low levels of differential gene regulation.

3.3.1.4 Gene Array Analysis

For data acquisition Arraygauge software was purchased from Fuji Film Corp. Arraygauge allows for independent analysis of spot intensity, background subtraction and membrane normalization. Values of spot intensity are acquired as exact pixel values or mean pixel value per spot area and sampled multiple times with in the spot perimeter. This software platform was chosen based on the ability to adjust multiple parameters, including membrane spot configuration, which allows alterations to the spotting format and accommodates differing array design including number of genes spotted on the array, spotting density, background, and normalization parameters.

Global background values are determined from regions of the membrane devoid of any hybridization signal or by analysis of internal controls placed on the membrane. In an initial series of experiments designed to test array dynamics, hybridization values with intensity > 10 were considered as positive signals. This value was calculated to ensure that hybridization values with low intensities were distinguished from the background with statistical significance >99.9% (Xu et al., 2001). Normalization of the arrays was based on the sum of background-subtracted signals from all of the genes on the membrane. This type of normalization procedure is possible due to the co-exposure of the control and treated membrane on the phosphor imaging screen where adjustments to signal intensity are made proportionally to both membranes.

Following completion of data acquisition, comparative analysis of treated to non-treated membranes was performed. Using the Arraygauge software, ratio values were determined for replicate spots on each membrane. Given the correct subtraction for background and normalization for global differences between hybridization (i.e. slight differences in probe concentration, membrane washing, and densitometry from phosphor image analysis), changes in gene expression were determined.

3.3.1.5 Array Reproducibility and Variance

In the initial phase of array development it is important establish specific components of array dynamics including array reproducibility, variance and sensitivity. To address these issues, a series of experiments were conducted using replicates of the 48-gene array. cDNA was isolated from ciprofibrate treated or control medaka liver and membranes hybridized with aliquots of the identical cDNA. These experiments were designed to determine consistency in membrane spotting, hybridization and washing techniques as well as establish a baseline of sensitivity and linear response. Results with the 48-gene array indicated a high degree of intra and inter-membrane consistency and sensitivity between experiments. Figure 3-1 demonstrates an example of independent trials for assessment technical reproducibility and variance. Analysis of variance using internal normalization of scatter response between log (1-4) suggest a variance ratio of no greater than 1.3. This data demonstrates that based solely gene expression, ratios greater than 1.3 can be scored as up or down regulation. Additionally these data illustrate that little technical variance occurs between membrane replications. In a similar set of experiments, biological reproducibility was examined by hybridizing replicate membranes with separately labeled probes from two independent (ciprofibrate treated) RNA sources. In all experiments equal amounts of starting RNA and probe concentration were added. In each case, data variance did not deviate above the 1.3 ratio level suggesting that both technical and biological variability are within an exceptable limit (standard practice for most microarray experiments is between 1.5 and 2). Additionally the 1:1 ratio line occurring at a 45° angle suggests that normalization between

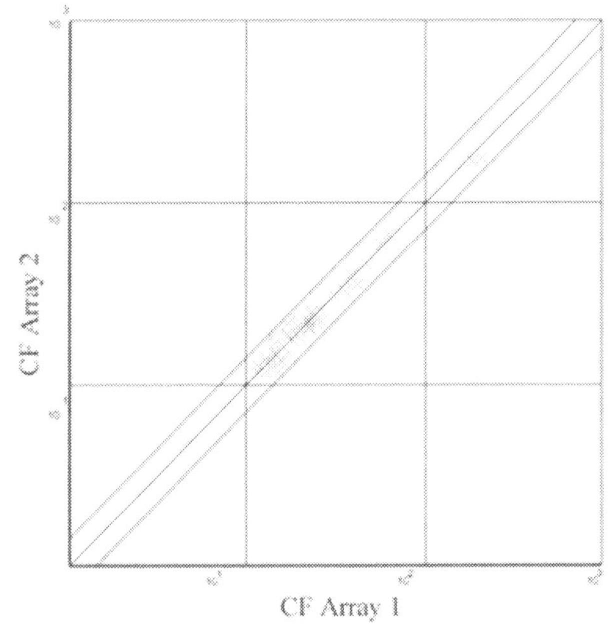

Figure 3-1. Log Scatter Plot of Two Independent Gene Arrays.
Analysis of the same liver sample illustrating membrane reproducibility and variance. Each point represents a gene with the X coordinate value as the gene expression level on one membrane and the Y coordinate value on the other membrane.

membrane intensities is proportionate. Most hybridization intensity values fell between $\log 10^1$ and 10^4 suggesting sensitivity range is at least three orders of magnitude. As stated above, background range values were between 6-7 pixel values (as determined in feature extraction using arraygauge software). Lowest positives scores and thus limit of sensitivity for these experiments were set at a pixel value of 10. The highest value recorded was approximately 10,000. Saturation for hybridization and/or phosphor imaging or data quantification (i.e. limits of Arraygauge software) was determined and shown not to influence the data set.

3.3.1.6 Preliminary Study with Multiple Fibrate Pharmaceuticals

Preliminary experiments were conducted to examine gene induction endpoints following acute exposure to members of the "fibrate" class of pharmaceuticals. Array validation for reproducibility and variance were determined by comparing two RNA pools of the same treatment. In this series of experiments, array hybridization was conducted using both control liver RNA and "fibrate treated" liver RNA to assess differences in gene expression following exposure. Ciprofibrate, clofibrate and clofibric acid were chosen as prototypic fibrate compounds based on their known molecular responses as peroxisome proliferators in other organisms. These additional pharmaceuticals were included to help differentiate and verify a ciprofibrate specific vs. class specific gene expression response. Total RNA was isolated from the liver of male medaka treated with 25 mg/kg body weight (i.p injection) of each pharmaceutical. cDNA probes were made according to the protocol described above (non-amplified) and hybridized to a replicate membrane.

Figure 3-2 illustrates data output as a scatter plot of gene expression response between replicate membranes hybridized with ^{32}P labeled probes from either control liver RNA vs ciprofibrate treated liver RNA or control liver RNA vs. clofibric acid treated liver RNA. For both experiments, array response was scattered over several orders of magnitude. Following background subtraction and data normalization, array ratio variance (treated/control) was set at 1.3. As in the initial experiments, gene expression responses greater than 1.3 are scored as up regulated and those less than 0.75 are scored as down regulated. Gene expression changes (both up and down regulation) were observed in both treatments suggesting differences between control and treated RNAs can be significantly detected with this platform. At the given dose (25 mg/kg) ratio values ranged between 0.3 and 3.0 indicating that while changes gene expression did occur they were not dramatic. From a survey of recent microarray data, it appears that the

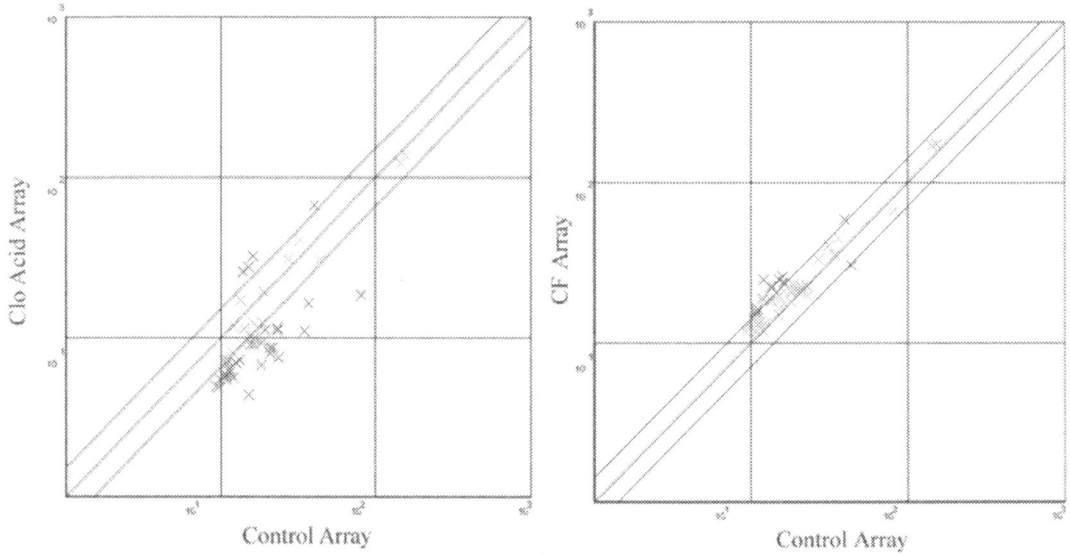

Figure 3-2. Scatter Plot of Gene Expression for Ciprofibrate vs. Control and Clofibric Acid vs. Control.
Each point represents a gene with the X coordinate value as the gene expression level on the control membrane and the Y coordinate value on the treated membrane.

majority of rank fold changes in gene expression (either up or down) occur between 2-5 fold. This is usually a result of slight changes in expression of constitutively expressed genes following treatments. More significant changes in gene expression are observed in highly inducible genes, for example Baker et al., 2003 has reported up to a 43.3-fold induction of Hydroxyacyl CoA dehydrogenase following exposure to clofibrate in rat liver. Studies with the 48-gene array ciprofibrate, clofibrate and clofibric acid exposure resulted in both up and down regulated genes for each compound tested. All treatments displayed similarity in responding gene sets including *precerebellin-like protein and c1q-like adipose protein. alpha tubulin* and *26S proteasome non-ATPase*.

Results with the clofibric acid exposure demonstrated an unusual number of genes that were down regulated. Clofibric acid is the acid analog of clofibrate and is used as an antihyperlipidemic. Unlike ciprofibrate and clofibrate, clofibric acid is not a peroxisome proliferating agent or a PPARα agonist. The fact that some gene changes are similar to all these pharmaceuticals then suggests that some gene responses in the medaka liver are not mediated by PPARα, suggesting an alternate mechanism of pharmaceutical action. This is not surprising given that several microarray experiments have recently been conducted with clofibrate in rat demonstrating multiple mechanisms of pharmaceutical action (Yadetie et al., 2001; Yamazaki et al., 2003). The log2 graph below illustrates changes in gene expression following clofibric acid exposure. When data is organized according to ratio value this type of graphical illustration is useful for demonstrating comparative expression changes on the array. In addition to genes mentioned earlier, treatment with clofibric acid induced *liver fatty acid binding protein, aldolase a, plasma glycoprotein, membrane translocase* and several down regulated genes associated with the acute phase response and immune function.

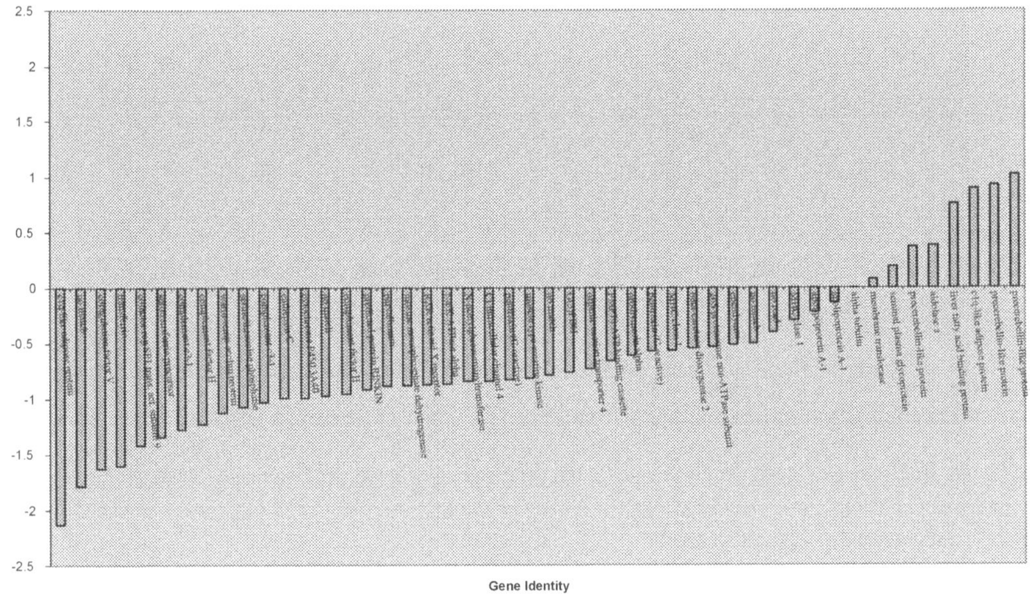

Figure 3-3. Log base 2 Plot for Gene Expression Analysis of Clofibric Acid Exposed Medaka Using Liver RNA on a 48 Gene Array.

3.3.2 Microarray Analysis of Fibrate Pharmaceuticals

To address the stated tasks of specific aim II, i.e. Establish linear range and sensitivity of membrane hybridization (dose response) and Evaluate acute dosing on gene induction endpoints (time to response), cDNA microarrays specific for medaka brain, liver, or testis were constructed. Brain, and liver, arrays contained 186, tissue specific genes each and testis arrays contained and 282 non-redundant testis specific genes. To assess intra-array variability, array controls were randomly spotted (25 ng/spot) on each membrane array. Array controls included: Arabidopsis Cab1 cDNA (6 spots/array), salmon sperm DNA (4 spots/array), and human β-actin (2 spots/array) (Spot Report-3 Array Validation System; Stratagene. Each gene was spotted in duplicate on each array for a total of 384 (Brain and liver) and 576 (testis) spots respectively. Duplicate spotting on each membrane and running each dose/time point as a biological replicate has provided ample data to assess the statistical significance of change in gene expression for each gene on each array. In some instances, a few genes were found to be expressed in multiple tissues and were placed on appropriate arrays. For example, PMP70 (a gene marker for Peroxisome Proliferation) was identified by SSH in the testis but was included on the medaka liver array were PMP70 is additionally known to be expressed.

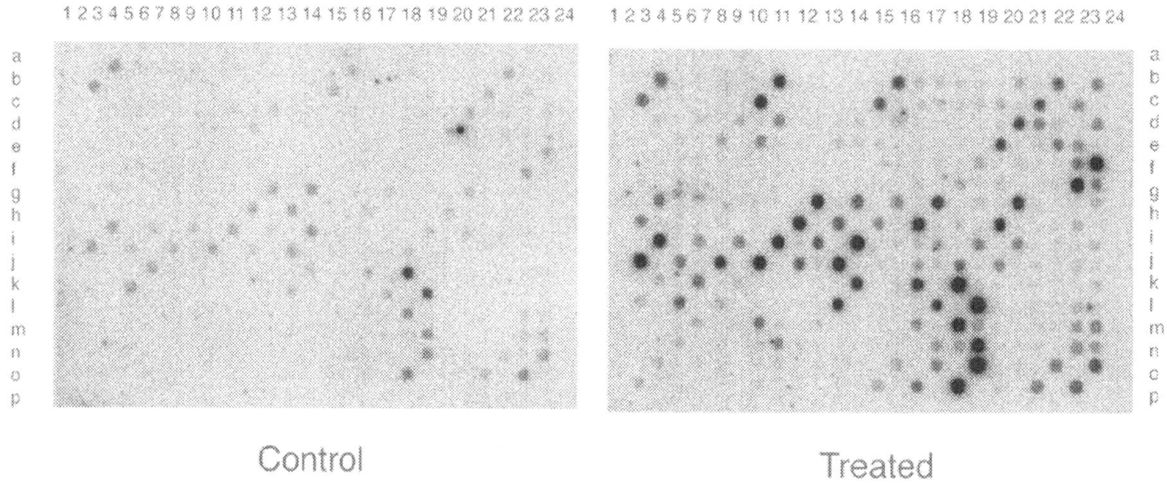

Figure 3-4. Medaka Liver Specific 192 Gene Array.
Array contains 186 non redundant medaka liver specific genes and 6 control genes. Each is spotted in duplicate on nylon membranes for a total of 384 gene targets (spots). All genes were spotted a final concentration of 25ng. The microarrays were hybridized with ^{32}P-labeled probes generated from control (DMSO treated) and Ciprofibrate treated (7-day exposure at 0.1 ug/L). Following hybridization membranes were washed (see material and methods above), and exposed to a phosphor screen for 72 h. Images were viewed using Arraygauge software.

3.3.2.1 Exposure Matrix

Exposure design incorporated time, dose, compound, and tissue specificity as illustrated in Table 3-1. This experimental design resulted in a total of 80 specific microarray hybridizations. All data has been acquired by exposing hybridized arrays to phosphorimaging screens for three days. Feature extraction was conducted with ArrayGauge v2.1 FugiFilm software allowing gene segmentation (i.e. defining individual spot size and diameter) spot intensity (total or mean), and membrane background (global or local).

Table 3-1. Dose Exposure Matrix.

DAY/DOSE	0.01mg/l	0.1 mg/l	1.0 mg/l	Control
Day-1	6 fish	6 fish	6 fish	6 fish
Day 3	6 fish	6 fish	6 fish	6 fish
Day 7	6 fish	6 fish	6 fish	6 fish

Compounds: Ciprofibrate and Clofibrate both members of the fibrate class of antihyperlipidemic agents. *Mechanism of action*: PPARα agonists.

Liver:12 conditions X 2 replications/condition 2 compounds = 48 membranes

Brain: 4 conditions (day 7 L, M, H and control) X 2 replicates/condition X 2 compounds =16 membranes

Testis: 4 conditions (day 7 L, M, H and control) X 2 replicates/condition X 2 compounds = 16 membranes

3.3.2.2 Array Analysis

As arrays were hybridized with ^{32}P-labeled cDNA probes, raw spot density data for each membrane were individually quantified. Raw image data were then exported as spreadsheets into Microsoft Excel, and coded by gene, treatment, treatment replicate, membrane replicate, and spot replicate. These hierarchical levels of identification were critical for appropriate statistical analyses outlined below. For each membrane, all raw spot intensity values were individually background adjusted for non-specific binding due to blocking agent (salmon sperm DNA) and normalized for hybridization differences among replicate membranes (*Arabidopsis Cab1*). Coded and normalized data for each sampling day (D1, D3, and D7) were then imported in SAS v9.1 (SAS Institute) and, within each treatment, tested for normality on a gene-by-gene basis using a UNIVARIATE procedure (α=0.05). As the sample size for each gene (N=4 spot values) was <2000, we based our assumptions of normality on the SAS-calculated Shapiro-Wilk statistic (W). This test statistic is strictly dependent on sample size, and small values of W (W<0.05) lead to rejection of the null hypothesis stating that the residuals are normally distributed. Based on these tests, it was determined that >75% of all possible gene-treatment conditions were normally distributed with equal variance, and was consequently, modeled using a logarithmic transformation typically used for statistical analysis of wide-scale gene microarray data. Least squared means of each biological and technical replicated were then calculated in SAS on a gene-by-gene basis using a mixed function linear model with a Bonferonni adjustment (Wolfinger et al., 2001). As array data were generated using a nested experimental design, the researchers used a mixed model to account for variability linked with each level of hierarchy. Unlike simple T-tests or standard linear models, this model considers both fixed- and random-effects parameters associated with known a priori-assigned variables (gene and treatment) and unknown random variables (treatment replicate, membrane replicate, and spot replicate), respectively, assumed to impact the variability of the data. Like fixed effects, random effects have probability distributions with means and variances. Thus, by modeling both fixed and random effects, the mixed model considers means, variances, and covariances within one model. In addition, for multiple comparisons of means, a Bonferonni correction was critical to control for high false-positive rates associated with large numbers of multiple comparisons when analyzing array data on a gene-by-gene basis. Significant treatment differences in array data were reported at or above the 90% confidence level (P<0.10) since this model was highly conservative. Treatment-to-control ratios for all genes were calculated based on least square means included within SAS output.

3.3.2.3 Identification of Dose and Time to Response Trends in Medaka Liver

The experimental design chosen incorporates several multifactorial relationships between dose, time, tissue, and compound. To begin to develop gene expression relationships the researchers employed hierarchical clustering and principal component analysis as an initial array diagnostic. Correlations between gene expression patterns can be investigated using a variety of techniques, including hierarchical clustering, *k*-means clustering, self-organizing maps, and PCA. Clusters identified using one technique can be marked and tracked through subsequent analyses. Cluster 3.1 and TreeView were used for organization and unsupervised analysis of the microarray data set.

The goal of gene expression data clustering is to group together genes and or treatments that exhibit similar expression patterns. It is useful for discovering the behavior of an expression matrix and aids in reducing the dimensionality of the data set, were all genes on an array are represented by a cluster of genes that behave similarly. In this case, both hierarchical clustering and principal component analysis were used as unsupervised approaches in initial data analysis. Both approaches are used in the absence of *a priori* information regarding treatment or gene annotation.

Figure 3-5 represents hierarchical clustering of clofibrate liver arrays. Clustering is based upon a correlation method that employs an agglomerative process that successively groups similar objects in a pair-wise iterative fashion until all data is included. The dendrogram is organized with treatments, both day and dose, across the X axis (top of the figure) and genes along the Y axis (left hand side). Branch lengths of the tree represent degrees of similarity between adjacent treatments or gene expression patterns. Hierarchical clustering was applied to gene expression profiles across fibrate treated animals at all three time points and all three doses. All genes represented in the cluster exhibited statistically significant expression alterations in at least one condition relative to the control. Clustering along the X axis demonstrates a distinctive relationship between gene expression and time of treatment. The resulting hierarchical ordering of the cluster tree suggests that the greatest correlation is observed between days 3 and 7 with decreasing correlation to day 1. This is consistent with observed ratio values were day one exhibits a higher frequency of down regulated genes compared to days 3 and 7 where gene expression ratios increase over time and with dose. These gene expression patterns reveal several key events following treatment including time dependent up-regulation, time dependent down regulation and transient up or down regulation.

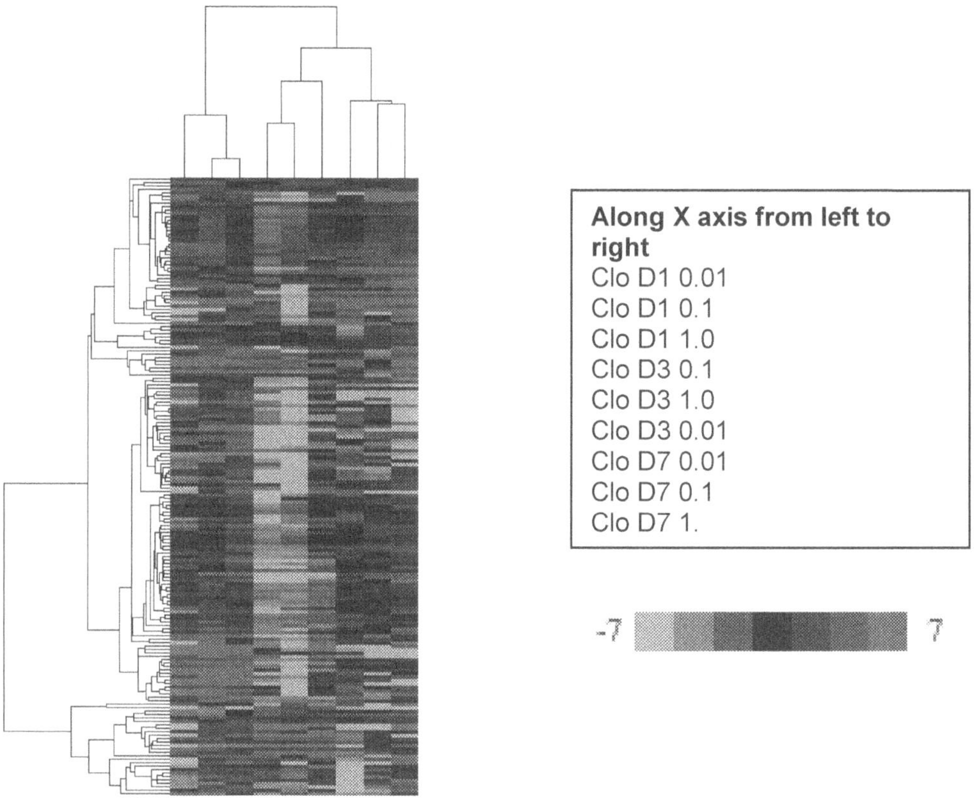

Figure 3-5. Hierarchical Clustering of Clofibrate.
Treatment (day and dose on the x axis and gene on the y axis). Red indicates up-regulation and green represents down regulation of gene expression.

When gene expression profiles for both ciprofibrate and clofibrate were clustered together (Figure 3-6), again time was the predominant factor for over all organization of the data. Within the clustering hierarchy however, days 3 and 7 exhibit dose related clustering particularly at the higher doses. For example CF D3 1.0, Clo D3 1.0 and CF D7 1.0 form a distinct cluster. Additionally dose related clustering is observed between CF D3 0.1 and CF D7 0.1. Theses data suggest that both time and dose are specific determinants of gene expression following fibrate exposure.

Clustering on the Y axis (grouping of genes) was highly divergent suggesting that gene expression profiles on an individual gene basis were highly independent. The researchers investigated however the hypothesis that genes associated with similar function would be regulated as a coordinated group. While in general it was found that gene expression was independent of gene function several clusters were observed that supported this general hypothesis specifically those genes associated with drug metabolism including CYP2P2, CYP1B, CYP1C and glucuronal transferases. Clustering of these genes was significant across time, were consistently down regulated, and did not exhibit any dose response relationships (see Figure A-4).

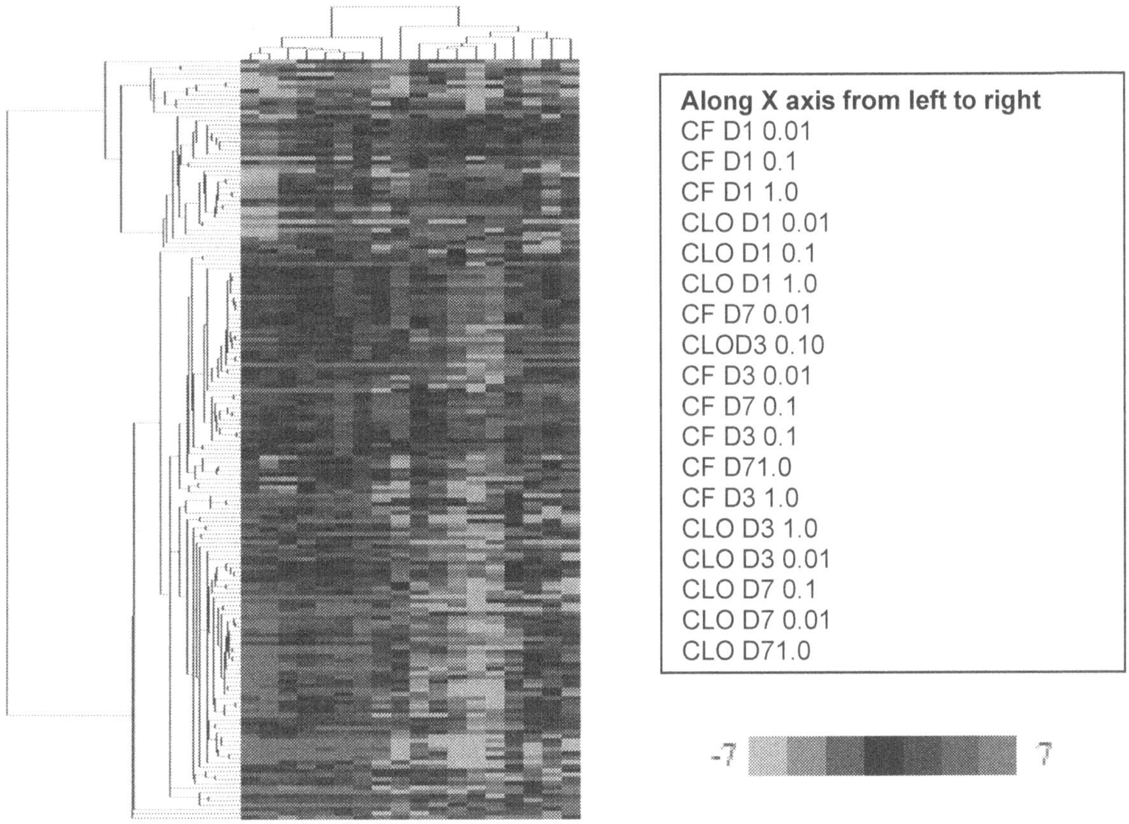

Figure 3-6. Hierarchical Clustering of Ciprofibrate and Clofibrate.
Treatment (day and dose on the X axis and gene on the Y axis). Red Indicates up-regulation and green represents down regulation of gene expression.

Principal component analysis (PCA) is an additional method used to visualize microarray data. PCA reduces the *dimensionality* of a data set and provides a visual mechanism to picture the structure of the data using as few variables as possible. The first principal component is the linear combination of the standardized original variables that has the greatest possible variance. Each subsequent principal component is the linear combination of the standardized original variables that has the greatest possible variance and is uncorrelated with all previously defined components.

As observed in hierarchical clustering, PCA (Figure 3-7) revealed a distinct separation between treatments with time and also demonstrated significant differences in relative proximity of samples with dose suggesting that in some instances a dose response may be evident. The first principal component accounted for ~ 68% of the variability in the data set and separated the samples similar to that observed with the clustering analysis i.e. a distinct separation with time. The second principal component which accounted for 15.5% of the variation, allowed the appreciation of the similarity in gene response between day 3 and day 7. The third principle component accounted for 6.3% of the data variability and clearly distinguishes gene expression on day 1 form both days 3 and 7. What is particularly interesting here is the pattern along the first principle component, where from left to right gene expression is separated according to time with both ciprofibrate and clofibrate. This level of organization for both the cluster analysis and PCA demonstrates the similarity in molecular response between these compounds with time as a major determinant. Factors affecting time to response include, duration of exposure, dose concentration, route of exposure and possibly handling of fish.

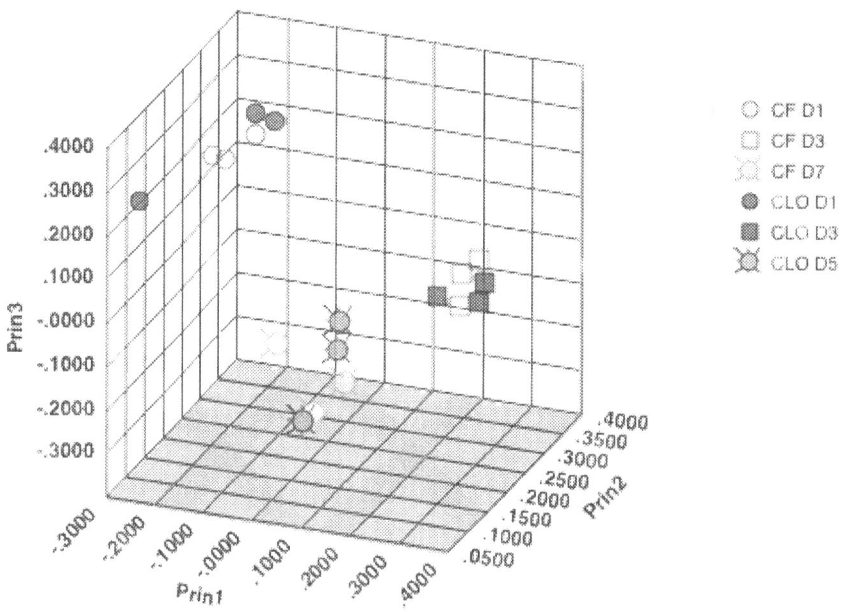

Figure 3-7. Principal Component Analysis. Ciprofibrate and Clofibrate days 1, 3, and 7.

While time to response drove the global organization of the data set, further analysis of gene expression changes within each day resulted in specific dose response relationships. Using a statistical permutation process data sets were examined for linear trends between each dose within each day. Figure 3-8 illustrates four distinct linear trends observed. The top left panel depicts a typical up regulation as a linear trend proceeds as a consistent increase in gene expression form control to low dose and upward to the high dose. Secondly, linear trends were identified in numerous genes as a consistent down regulation. Two other types of linear trends were additionally observed. These represented mirror images of the "normal" up or down regulation usually observed. The first depicts gene expression profiles were exposure to the low dose results in a significant down regulation in comparison to the non-treated control. With increasing dose gene expression increases back toward or in some instances past the original control value. In he fourth type linear trend observed, low dose exposure results in a significant increase in gene expression above control values. With increasing dose however gene expression changes decrease to original control levels or in some instances below control values appearing as a down regulated response. The last two types of linear dose responses are difficult to explain however occur often within the data sets and represent significant gene expression changes for ciprofibrate and clofibrate exposed liver. Non-linear dose response changes were additionally evident with both liver data sets. These gene expression relationships were more indicative of a non-monotonic, biphasic "U" or inverted "U" shaped dose response where the middle dose is distinctly different from the lower and higher dose. This type of dose response is consistently observed with endocrine disrupting compounds and is a subject of study. From a biomarker perspective we are more interested in genes that demonstrate a clear and distinctive dose response relationship. Thus an estimation of exposure can be determined based on the degree of expression. Alternatively a qualitative determination of exposure can be made based on observation of a specific gene expression profile. In keeping with this design we have compared gene expression response between both ciprofibrate and clofibrate for gene sets that are either consistently up or down regulated across dose.

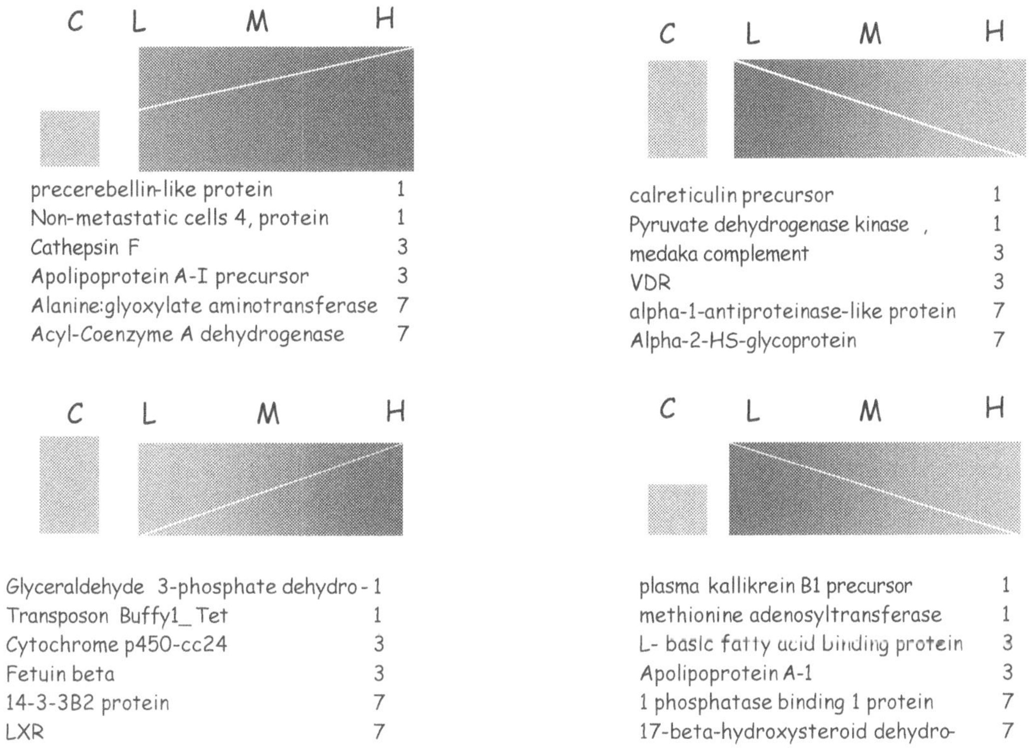

Figure 3-8. Examples of Linear Trends in Dose-Response Per Day.
Linear trend analysis was conducted using the "R" statistical analysis package. Gene names are given below each linear trend type as examples. Numbers following each gene name is the day in which this trend was observed.

3.3.2.4 Gene Expression Similarities in Medaka Liver

Given that the two fibrate-based pharmaceutical compounds share a similar mechanism of action (i.e. PPARα agonists) it was hypothesized that similar gene expression profiles would be observed. This hypothesis is the basis on the identification of class specific markers that can discern exposure to select compounds. The assumptions to this approach however are that both compounds display similar binding kinetics to the PPAR receptor (i.e.similar binding affinities) and if any, share similar non-receptor mediated mechanisms- effects. As such, Table 3-2 lists a set of genes that exhibit similar gene expression profile for both ciprofibrate and chlofibrate. The data shown illustrates that within both dose and time (here 1.0 ppm, day 7) a similar set of genes are up regulated. While the degree of gene expression change is not exact the general trend is a coordinated response to both compounds at this dose and time.

To determine the consistency of gene expression across doses, gene sets were identified where individual genes were either up regulated 1.3 fold or down regulated 0.75 fold across each dose for both compounds. Given the small degree of variation in gene response we set a cut off value were any one gene was considered consistently up regulated if it was present in five out of the six treatment in a given day (i.e. D1- 0.01, 0.1, 1.0 CF and D1-.01, 0.1, 1.0 Clo). In this analysis it was demonstrated that within each day, there are specific sets of genes that reflect unique expression patterns (up and down) for the fibrate drugs. Gene sets differed significantly by day with little overlap in expression profiles. As an example, the gene set for Day 1 all doses is shown in Table 3-3. Fifteen genes were consistently up regulated between ciprofibrate (medium and high dose) and clofibrate (low, medium and high dose). Similarly 41 genes were

Table 3-2. Similarity in Gene Response: CF & Clo Day 7, 1ppm.

Gene	Ratio CF	Gene	Ratio CLO
1-Cys Peroxiredoxin	1.721230947	1-Cys Peroxiredoxin	2.852438505
Annexin max3	1.740389036	Annexin max3	2.896474951
Aquaporin	1.582632843	Aquaporin	2.74065563
Bile salt-activated lipase	3.403884558	Bile salt-activated lipase	4.704982101
Calmodulin	5.150123455	Calmodulin	4.954013434
CDC-like kinase 4 (Clk4) protein	1.790344797	CDC-like kinase 4 (Clk4) protein	3.334565722
Cytochrome P450 1B (CYP1B)	2.772584356	Cytochrome P450 1B (CYP1B)	2.546309841
HNF4	2.071870251	HNF4	2.867256864
Hypothetical protein LOC402870 (Danio rerio)	2.374479055	Hypothetical protein LOC402870 (Danio rerio)	3.916693181
Kallikrein	2.253008773	Kallikrein	2.205702966
Leukocyte surface protein precursor	1.685995345	Leukocyte surface protein precursor	3.125487785
Macrophage mannose receptor	1.796126744	Macrophage mannose receptor	3.226390277
Mitogen-activated protein kinase phosphatase 2	2.71860536	Mitogen-activated protein kinase phosphatase 2	2.80726177
Myo-inositol oxygenase (MIOX)	1.786800101	Myo-inositol oxygenase (MIOX)	2.931642208
P5-pending protein/protein disulfide isomerase-related protein	2.067167699	P5-pending protein/protein disulfide isomerase-related protein	2.099931135
Reticulon 4 receptor-like 2b	3.619972732	Reticulon 4 receptor-like 2b	4.711646767
RIKEN cDNA 2410046H15 (LOC306030)	3.206721879	RIKEN cDNA 2410046H15 (LOC306030)	5.230158787
RXR, retinoid X receptor	2.127132579	RXR, retinoid X receptor	2.119022602
Thioredoxin domain containing 4 (endoplasmic reticulum)	3.585272524	Thioredoxin domain containing 4 (endoplasmic reticulum)	4.830854317
THR	1.907162163	THR	4.563409715
Unknown (protein for MGC:55636) [Danio rerio]	1.947388381	Unknown (protein for MGC:55636) [Danio rerio]	3.211225993
Wu:fk63d09	4.503392823	Wu:fk63d09	4.351212751

consistently down regulated across five of six doses. In this type of analysis day 1 had the most down-regulated genes and day 3 the most up-regulated (see Appendix A-5). Thus while there appears to be consistency in gene expression across dose, differing durations of exposure result in unique gene expression profiles by day.

Given the consistency of the gene response, expression profiles can be extended as biomarkers of exposure. While this type of analysis will not provide information regarding the degree of exposure (if such a graded response exists) it will provide information that exposure has occurred. This may be typical of gene array data in general. As more microarray experiments are conducted, a general consensus is there limited sensitivity for quantitative analytical analysis. In fact while gene arrays maybe highly beneficial for establishing profiles of expression (which will be particularly useful for mixtures) more exact quantitative analysis may be obtained with quantitative polymerase chain reaction (qPCR) or specific bioanalytical assays such as those produced and validated for endocrine disrupting compounds (ex. YES or ESCREEN assays).

As an additional level of scrutiny we have applied a stringent requirement that individual genes demonstrate a statistically significant responses. As array data were generated using a nested experimental design, and a mixed model to account for variability linked with each level of hierarchy. Unlike simple T-tests or standard linear models, this model considers both fixed- and random-effects parameters associated with known a priori-assigned variables (gene and treatment) and unknown random variables (treatment replicate, membrane replicate, and spot replicate), respectively, assumed to impact the variability of the data. Like fixed effects, random effects have probability distributions with means and variances.

Table 3-3. Gene Expression Similarities Across Dose (0.01, 0.1 &1.0 Ppm) and Compound (CF & Clo), at Day 1.
Left Hand Side Up Regulated Genes 5/6 Treatments. Right Hand Side Down Regulated Genes 5/6 Treatments

CF/CLO Day 1 Dose 0.01-1.0 UP	CF/CLO Day1 Dose 0.01-1.0 DOWN
1-Cys Peroxiredoxin	17-beta-hydroxysteroid dehydrogenase type 4
acetyl-CoA carboxylase	Annexin max3
Aldolase A	Arrestin domain containing 2
Coagulation factor XIII b subunit precursor	Bile salt-activated lipase
Hepatocyte growth factor-like 1 (hgfl1)	Cathepsin B precursor
Hypothetical protein LOC402870 (Danio rerio)	CDC-like kinase 4 (Clk4) protein
Hypothetical protein MGC63938 (Danio rerio)	CG4071-PA
Non-metastatic cells 4, protein	Clone DKEY-202E22 in linkage group 21
pentraxin [Cyprinus carpio]	Cone potassium-dependent sodium-calcium exchanger
Phospholipid scramblase Zic1	CXCR3 gene for CXC chemokine receptor
Properdin factor, complement	Cytochrome P450 1A (CYP1A)
RXR, retinoid X receptor	Cytochrome P450 1B (CYP1B)
Stromal cell derived factor 1	Cytochrome P450 1C2
UDP-glucoronosyl transferase	Cytochrome P450 2P2 (CYP2P2)
Wu:fk63d09	DKFZP434K046 protein
	DnaJ (Hsp40)
	Fetuin beta
	Glucose-6-phosphatase
	Glyceraldehyde 3-phosphate dehydrogenase
	Inter-alpha trypsin inhibitor, heavy chain 3
	Kallikrein
	Leukocyte surface protein precursor
	LXR
	Membrane-associated protein HEM-1
	MHC Class I
	P5-pending protein/protein disulfide isomerase-related protein-parvalbumin
	PTPa and cGnRH-II
	Putative disease resistance protein
	Reticulon 4 receptor-like 2b
	RIKEN cDNA 2410046H15 (LOC306030)
	Syndecan binding protein (SDCBP)
	THR
	TPA: replicase/helicase/endonuclease
	Transcription factor nrf2
	Transposon Buffy1_Tet
	Tyrosine aminotransferase
	Unknown (protein for MGC:56244) [Danio rerio]
	VDR
	vitellogenin 1
	zt1.9 microsatellite sequence

Thus, by modeling both fixed and random effects, the mixed model considers means, variances, and covariances within one model. In addition, for multiple comparisons of means, a Bonferonni correction was critical to control for high false-positive rates associated with large numbers of multiple comparisons when analyzing array data on a gene-by-gene basis. Significant treatment differences in array data were reported at or above the 90% confidence level ($P<0.10$) since this model was highly conservative. Treatment-to-control ratios for all genes were calculated based on least square means included within SAS output.

In Figure 3-9 the researchers demonstrate that within each treatment (day and dose) a significant number of genes are statistically different from matched controls. Use of the mixed model increase the assurance that gene changes accurately reflect gene expression changes due to exposure rather than a random event or type one statistical errors. While initial analysis was

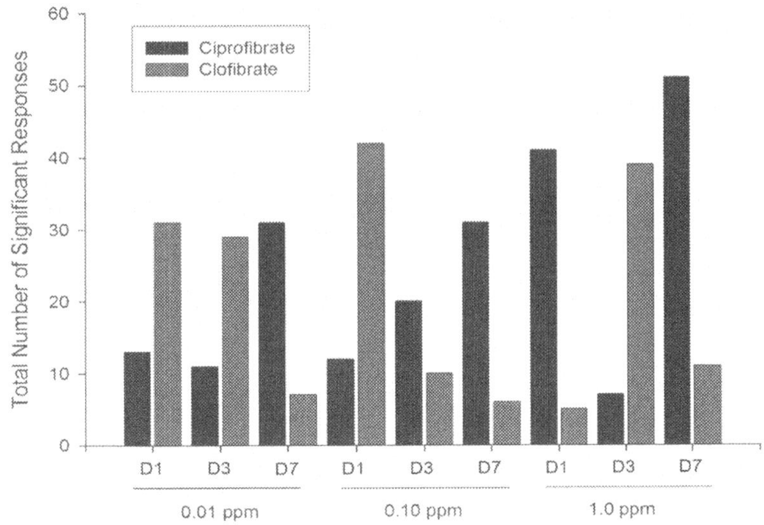

Figure 3-9. Total Number of Significant Responses.
Significance is determined between treated and control values for individual genes on the liver array using the SAS mixed function model as described in Section 3.3.2.2. Significance is presented as number of significant genes per day at a given dose.

based on gene expression changes between 1.3 fold up and 0.75 fold down, by applying a statistical significance to our data we increase the stringency for identifying truly differentially regulated gene sets. This analysis results in two specific changes to the expression profile, 1) compression of the data set by elimination of genes that while highly up or down regulated are not statistically significant and 2) expansion of the data set by addition of genes with high significance. While individual genes in the latter category may not be highly up or down regulated they may still play an important part in the over all expression profile.

An additional requirement for a well-defined expression profile is the dose response relationship. While this data set does not exhibit global organizations based on dose, several dose response relationships are identified within a given day and compound. Thus using three general requirements; 1) high to moderate gene expression change from control value, 2) statistically significant difference between control and treated gene expression values and 3) dose response relationships, the researchers identify ideal parameters to establish a class specific expression profile. The major caveat to these requirements are however, that the profile must be distinctive to the compounds examined and little overlap should occur in gene sets with other pharmaceutical classes. While not an aim in this study, initial comparisons of fibrate gene expression profiles to other on-going microarray studies in the medaka to 17β-estradiol and ethinyl estradiol are presented in Figure 3-10. Hierarchical clustering of fibrates and estrogens in liver arrays from high dose treatment (1.0 ppm) for seven days is clearly organized into two distinct expression groups with CF and Clo forming one branch of the tree and E2 and EE forming the other branch along the upper X axis. Branch lengths of the tree represent degrees of similarity between adjacent treatments or gene expression patterns. While CF and Clo exhibit similar gene expression patterns, similarity within the estrogen treatments is stronger as evidenced by the shorter branching distance from the root of the tree. Correlation analysis demonstrates a high degree of similarity within each pharmaceutical class (fibrates R^2, CF vs. Clo ~0.65, estrogens R^2, E2 vs. EE ~0.92) and little similarity between classes and (fibrates-estrogens R^2, CF vs. E2 ~0.05). Principal component analysis, additionally, suggests that the two pharmaceutical classes form distinct clusters, demonstrating highly divergent gene expression patterns (data not shown). However when applying our three criteria for developing and analysis of gene expression profiles, i.e. 1) high to moderate gene expression change from control value, 2) statistically significant difference between control and treated gene expression values and 3) dose response relationships, closer examination suggested that there are some overlapping changes in gene expression. As an example of this we have selected gene sets that are

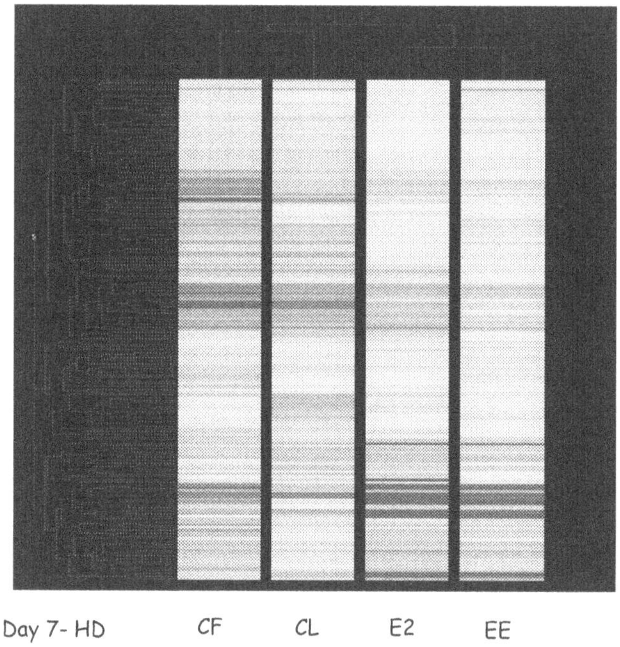

Figure 3-10. Hierarchical Clustering of Clofibrate, Ciprofibrate, 17β-estradiol and Ethinyl Estradiol.
Treatments, day and dose on the X axis and gene name, cluster on the Y axis. Red indicates up-regulation and green represents down regulation of gene expression.

representative of either fibrate (common between CF and Clo) or estrogen (common between E2 and EE) exposures at 1ppm for seven days Table 3-4. Listed are up regulated gene expression changes for both pharmaceutical classes. While at first glance gene expression appears to be highly specific, closer examination reveals several individual genes that are co-expressed (at this time and dose) between the two series of experiments.

In Table 3-4 those genes highlighted represent genes common to both fibrate and estrogen exposed fish. Interestingly, two genes thought to be specific for fibrates are also regulated by estrogens including RXR and PMP70. In this instance using a single gene biomarker approach to evaluate exposure, would produce a false positive for fibrate exposure. Alternatively we can focus on the patterns of gene expression or expression profiles and observe distinct differences in gene regulation. Within these patterns we may find induction/repression of specific genes that are unique to a chemical class such as vitellogenin for the estrogens or CYP1A induction for co-planer aromatic hydrocarbons. These telltale biomarkers will aid in determining exposures however may not be sufficient for an unequivocal determination of exposure. Additionally if we examine gene annotations, it becomes obvious that specific chemical classes will elicit defined metabolic pathways based upon their mechanism of action. As described in Chapter 2.0, fibrate drugs are PPARα agonist and thus result in expression of a defined set of genes and pathways associated with lipid metabolism and transport, amino acid metabolism and others that can be exploited for profile analysis.

Table 3-4. Comparison of Gene Expression Changes Between Fibrates and Estrogens 1ppm Day-7.

Gene Fibrates D7 HD	Ratio	Gene Estrogens D7 HD	Ratio
P5-pending protein/protein disulfide isomerase-related protein	7.13	vitellogenin 1	11.65
Membrane guanylyl cyclase OlGC2	7.05	CDC-like kinase 4 (Clk4) protein	8.01
Phosphoenolpyruvate carboxykinase, cytosolic [GTP]	6.45	P5-pending protein/protein disulfide isomerase-related protein	7.72
Leukocyte surface protein precursor	5.82	Nuclear receptor-binding factor 1	7.56
Unknown (protein for MGC:55636) [Danio rerio]	5.2	Tagln2-prov protein	7.44
Arrestin domain containing 2	5.16	Calmodulin	7.31
Nuclear receptor-binding factor 1	4.24	PMP70 ATP binding Cassette	7.22
High choriolytic enzyme 1 precursor	3.85	transposase/von Hippel-Lindau binding protein 1	7.2
Aquaporin	3.76	Transposon Buffy1_Tet	7.18
Wu:fk63d09	3.69	Wu:fk63d09	7.05
Gonadotropin-releasing hormone receptor	3.67	High choriolytic enzyme 1 precursor	7.04
RXR, retinoid X receptor	3.27	Annexin max3	6.75
acetyl-CoA carboxylase	3.2	Cytochrome P450 2P2 (CYP2P2)	6.43
Ferritin	3.17	RXR, retinoid X receptor	2.87
Aldehyde dehydrogenase 12	2.92	Cytochrome P450 1B (CYP1B)	2.5
pentraxin [Cyprinus carpio]	2.86	Gonadotropin-releasing hormone receptor	1.96
RIKEN cDNA 1600013K19	2.26	Kallikrein	1.7
alanine:glyoxylate aminotransferase	2.13	pentraxin [Cyprinus carpio]	1.65
Etr1	2.06	L-SF precursor	1.56
acyl-Coenzyme A oxidase 3, pristanoyl	2.06	TPA: replicase/helicase/endonuclease	1.53
Cytochrome p450-cc24	2	Arrestin domain containing 2	1.52
Transposon Buffy1_Tet	2	Cytochrome P450 1C2	1.51
parvalbumin	1.95	Coxsackie virus and adenovirus receptor	1.49
Eef2-prov protein	1.89	Inter-alpha trypsin inhibitor, heavy chain 3	1.46
Complement C5-2	1.87	Syndecan binding protein (SDCBP)	1.46
Elastase 1 precursor	1.87		
14-3-3B2 protein	1.87		
Properdin factor, complement	1.77		
PMP70 ATP binding Cassette	1.76		
complement factor H-related protein	1.75		
Membrane-associated protein HEM-1	1.75		
Sialic acid binding Ig-like lectin 6, isoform b	1.56		

Note: shaded areas similar between Fibrates and Estrogens

To illustrate this point, a pathway analysis was conducted (using the Ingenuity software package) on gene expression profiles obtained in this study. This software enables input of specific gene sets (in this case differentially regulated following fibrate exposure). The data is then compared to a vast, computable, biological knowledge base that computes a set of relevant pathways and networks. Pathways and networks are created from millions of individually modeled relationships between proteins, genes, complexes, cells, tissues, drugs, and diseases. Using this knowledge base, researchers can take a bioinformatics approach to understanding mechanisms of action of fibrates in teleost organisms. This approach allows investigators to systematically analyze experimental data in the context of known biological pathways, mechanisms of toxicity, and biomarker identification. Table 3-5 illustrates results of the pathway analysis in this study. Clearly defined are biochemical pathways associated with PPARα agonist and other cellular mechanisms influenced by fibrate drugs. These data suggest that fibrates function via similar cellular mechanisms in medaka to that observed with mammals. Given the vast knowledge base of PPARα agonists in mammals, these gene expression studies may help facilitate extrapolation of mechanistic responses across species. Ultimately this will help in the quest to develop bio-analytical approaches for determination of chemical compounds in reuse waters, drinking waters or STP effluents.

Table 3-5. Ingenuity Pathway Analysis.

Gene Stets from Fibrate Exposed Medaka Were Compiled and Analyzed for Pathway Identification Using Ingenuity Pathway Analysis Software Package. Human and Rodent Accession Numbers Were Substituted for Medaka Homologs to Enable Gene Identifications.

Biochemical Pathway	Sig α 0.05	Gene Symbol
Gene Regulation By PPAR(alpha)	(1.0e-15) (5)	FABP1, APOA1, Ins1, NR1H3, HSD17B4
KEGG Fatty Acid Metabolism		
Glycolysis / Gluconeogenesis	(1.5e-08) (8)	ALDOA, ALDOB, G6PC, ALDH1A2, 3A1, 3A2, 3B1, 3B2
Extrinsic Prothrombin Activation Pathway	(4.4e-05) (3)	F2, F10, FGA
Intrinsic Prothrombin Activation Pathway	(4.4e-08) (5)	KLKB1, F2, F10, F11, FGA
Classical Complement Pathway	(5.8e-11) (6)	C2, C4A, C4B, C3, C1R, C9
Complement Pathway	(7e-12) (7)	C2, C4A, C4B, C3, C1R, C9, BF
Blood Coagulation: BP	(2.7e-06) (6)	F11, F13B, F10, F2, FGA, FGG
Complement Activation: BP	(7.9e-09) (6)	C2, C4A, C4B, BF, HF1, HFL1
Alternative Complement Pathway	(2.3e-05) (3)	C9, C3, BF
Lectin Induced Complement Pathway	(8.2e-09) (5)	C2, C4A, C4B, C3, C9
Alcohol Dehydrogenase/biosynthesis	(0.00073) (3)	CYP1A1, ALDH3B1, ALDH3B2
Alcohol Dehydrogenase/genetics	(8.2e-05) (5)	ALDH3A2, ALDH3B1, ALDH3B2, C3, CYP1A1
Ethanol/pharmacokinetics	(0.00021) (3)	ALDH3A2, ALDH3B1, ALDH3B2
Polyisoprenyl Phosphates	(0.00065) (4)	C1R, HF1, C2, ACACB
Tangier Disease/metabolism	(0.00045) (3)	C2, ACACB, C1R
Acute Myocardial Infarction	(7.6e-05) (3)	F2, F10, FGA
Aldehyde Dehydrogenase Activity: MF	(1.3e-08) (5)	ALDH1A2, ALDH3A1, ALDH3A2, ALDH3B1, ALDH3B2
Aldehyde Oxidoreductase Activity,	(1.2e-07) (5)	ALDH1A2, ALDH3A1, ALDH3A2, ALDH3B1, ALDH3B2
Histidine Metabolism	(5.9e-06) (5)	ALDH1A2, ALDH3A1, ALDH3A2, ALDH3B1, ALDH3B2
Phenylalanine Metabolism	(0.0009) (3)	ALDH3A1, ALDH3B1, ALDH3B2
Propanoate Metabolism	(0.00018) (4)	ALDH1A2, ALDH3A2, ACACA, ACACB
Pyruvate Metabolism	(0.00025) (4)	ALDH1A2, ALDH3A2, ACACA, ACACB
Tryptophan Metabolism	(0.021) (3)	CYP1A1, ALDH1A2, ALDH3A2
Tyrosine Metabolism	(0.0041) (3)	ALDH3A1, ALDH3B1, ALDH3B2

3.3.2.5 Gene Response in Brain and Testis

Secondary to liver, brain and testis were identified as potential targets for fibrate drugs. The researchers thus included these tissues as part of a multi-organ toxicogenomic study. Analysis of testis and brain was conducted on day 7 for low, medium, and high dose where significant gene changes were identified for both ciprofibrate and clofibrate in these tissues. As with liver, gene sets were developed by identifying those genes consistently up or down regulated. These data sets were generated by selecting gene expression data common between five of the six treatment groups (eg. 0.01, 0.1, 1.0 CF, and 0.01, 0.1, 1.0 Clo) as demonstrated in Appendix A-7 and A-8. Gene expression ratios for brain and testis were highly similar for each gene within the six treatments groups.

In brain, ciprofibrate and clofibrate induced a dose-dependent increase in the number of significantly responsive genes ($P<0.1$). Significantly responsive genes were predominantly up regulated (some down) and ~40% of these responsive genes clustered with processes involved in brain development. Moreover, adenosine deaminase, mitogen activated protein kinase phosphotase, vacuolar protein sorting 26, and Zic family member—genes involved in neurogenesis, energy homeostasis, and protein transport—were consistently up-regulated >1.3-fold in at least five of six treatment groups. Among all treatments, only 21% (39/186) and 15 % (26/186) of the total genes screened were significantly responsive in ciprofibrate and clofibrate-treated adult medaka brains, indicating that statistical analyses were robust enough to detect mRNA differences due to pharmaceutical exposure.

In testis ciprofibrate and clofibrate induced a dose-dependent increase in the number of significantly responsive genes ($P<0.1$); these however were down regulated in comparison to the brain. While the list of commonly expressed genes in Appendix A-8 exhibits numerous up regulated genes, remarkably most of these were not statistically significant. Thus in this instance, screening for statistical significance compressed the expression profile.

Table 3-6. Statistically Significant Gene Response for Brain and Testis.

CF Brain					CF Testis				
Dose	Day	TRT/CTL <1	TRT/CTL >1	Total # Genes	Dose	Day	TRT/CTL <1	TRT/CTL >1	Total # Genes
0.001	7	1	6	7	0.001	7	30	2	32
0.1	7	0	15	15	0.1	7	32	4	36
1	7	0	17	17	1	7	33	4	37
CLO Brain					**CLO Testis**				
Dose	Day	TRT/CTL <1	TRT/CTL >1	Total # Genes	Dose	Day	TRT/CTL <1	TRT/CTL >1	Total # Genes
0.01	7	2	8	10	0.001	7	21	16	37
0.1	7	2	10	12	0.1	7	27	5	32
1	7	3	3	6	1	7	29	9	38

3.4 Concluding Remarks

Chapter 3.0 addresses tasks stated in Aim II which attempts to characterize differentially expressed cDNAs and gene expression profiles as markers of pharmaceutical exposure and effect by membrane hybridization. The proposed tasks were:

- Task 1. Establish linear range and sensitivity of membrane hybridization
- Task 2. Evaluate acute dosing on gene induction endpoints
- Task 3. Sequence and identify cDNAs by sequence homology analysis
- Task 4. Evaluate feasibility and sensitivity of array analysis to reuse water matrices.

Chapter 3.0 specifically addresses tasks 1-2. (See Chapter 2.0 for task 3 and Chapter 4.0 for task 4).

Subsequent to development of a ciprofibrate responsive EST library, gene sequences were used to generate spotted cDNA microarrays and tested for usefulness in expression profiling experiments. In initial studies specific array parameters and dynamics were established including spotting of DNA targets on membranes, non-amplified versus amplified probe production, software for image analyses, and development of internal standards for normalization. Each of these factors is important and has contributed significantly to the development of a microarray that can be used to establish a set of gene markers for assessment of fibrate pharmaceutical exposures in medaka. In Section 3.3 a mock 48-gene array was used to establish array reproducibility and variance and generate preliminary data on gene expression for multiple members of the fibrate class including ciprofibrate, clofibrate and clofibric acid. In Section 3.4 a more thorough analysis of gene expression was conducted using an expanded set of arrays which incorporated a larger set of medaka specific genes (from additional subtractions) totaling 186 genes on the liver array, 186 genes on the brain array and 292 genes on the testis array. Using a standard reference array design, time and dose profiles were produced for pharmaceutical agonists. Using replicate arrays to assess intra- and inter-array variability, as well as rigorous statistical approaches for array data analysis, the researchers successfully validated the utility of these custom arrays for semi-quantitative evaluation of transcript levels in adult medaka. Data was acquired using standard feature extraction software and hierarchical levels of identification were determined for, treatment, (dose and time), treatment replicate (biological), array replicate (technical) and spot replicate (also technical), using appropriate statistical analyses.

Using unsupervised analysis including hierarchical clustering and principal component analysis, trends in gene expression were observed including time and dose relationships. General results of this study suggest that gene expression was more a function of time than dose however specific dose response patterns were observed within a given day. As stated it is hypothesized that these results are due to the high dose range used in this study were gene response maybe saturating. A primary goal of this study however was to identify genes that were induced or repressed following exposure. While a graded gene response may not have been observed, consistency in gene expression changes were established suggesting that exposure to fibrates results in a defined set of molecular events. Comparisons between ciprofibrate and clofibrate across time and dose demonstrated that expression profiles can be established for this class of compounds and used for exposure assessment. One caveat to this type of analysis however is that other classes of pharmaceuticals do not produce a similar gene expression profile. For comparative purposes hierarchical clustering, PCA and correlation analysis was conducted between ciprofibrate and clofibrate, 17β estradiol and ethinyl estradiol two distinct pharmaceutical classes. Results of this analysis suggest that global gene expression profiles are highly unique for each pharmaceutical class. However examination of individual genes exhibited some degree of overlap further supporting the need to examine global gene expression profiles compared to expression of a few distinct genes. Within this analysis the researchers have developed several distinct criteria 1) genes must be reasonably up or down regulated in comparison to control, 2) gene expression changes (within the profile used) must exhibit statistical significance, 3) (ideally) gene expression change should exhibit dose response relationships, and 4) gene expression changes should be consistent across members of pharmaceutical class. While the data sets generated in this study meet most of these criteria, there is room for improvement. The researchers envision this first set of arrays as development of a training data set (i.e. the gene expression profile), which will be subsequently improved upon and challenged with increasing repetition. In subsequent experiments data should be acquired from exposures at lower concentrations, which may result in established dose response relationships. This would additionally relate gene expression to environmentally relevant concentrations. Secondly increasing both the biological and technical replication of the arrays will aid in generating more statistical relevance, validating gene expression changes. Third, additional members of the pharmaceutical class should be compared to the original training set and last analysis between pharmaceutical classes should be conducted.

Through this analysis the researchers have generated a specific set of gene expression profiles believed useful as a bioanalytical tool for contaminant assessment in reuse waters and other aquatic environments. As with any new tool the use of microarray analysis will take time to become an established assessment method. These studies were designed as proof of principle for such an application. The water quality community is well poised to evaluate the feasibility of array analysis and its application toward assessment of reuse and wastewater quality.

CHAPTER 4.0

MICROARRAY ANALYSIS IN REUSE WATER

4.1 Introduction

The aim of this task was to evaluate feasibility and sensitivity of using microarray analysis for detection of fibrate based pharmaceuticals in reuse water. As such, the researchers employed previously established bioassay and microarray protocols to test the applicability of this approach to assess contaminant activity in real water matrices. In this task, reuse water was collected from a single wastewater treatment facility in Raleigh, North Carolina. The Raleigh facility uses an advanced activated sludge tertiary system with denitrification and UV disinfection. This facility was chosen based on standard water treatment practice and was recommend by both the PCS committee and collaborators on this project. Prior to analysis, water quality parameters including TOC, BOD, pH, total nitrogen, alkalinity, and turbidity were determined to assess if water conditions were appropriate for fish exposures. Reuse waters were subsequently spiked with clofibrate as a model compound following protocols established in earlier phase of the project. 1ppb clofibrate was chosen as a low concentration to provide a dose that approximated environmental relevance. A second concentration of 10 ppb was added to an additional group of fish to anchor the studies and allow comparisons to previous studies conducted in laboratory water (see Chapters 2.0 and 3.0). Given that original studies suggested that time was a significant factor governing gene expression the researchers chose to assess gene expression changes following acute (24 hrs) and sub-acute (72 hrs) exposures. To ensure that this approach produced a data set with appropriate statistical significance, the experimental design incorporated three biological replicates comprised of three fish/replicate, each exposed in independent chambers. At each time point, livers of individual fish from each biological replicate were collected and pooled. Each biological replicate was then hybridized to two separated liver specific microarrays resulting in two technical replicates for a total of six complimentary arrays/treatment. While all treatments were conducted in reuse water, an additional control was conducted in fish rearing media (laboratory water) to enable comparisons to reuse water controls. This experimental design has enabled assessment of effects associated with the composition of reuse water itself (see matrix below). Studies were conducted with the orange red strain of medaka (six months in age) in order to make direct comparisons to the training data set produced with laboratory water.

4.2 Matrials and Methods

4.2.1 Exposures

One hundred adult male medaka (six-months old) were selected at random from the Duke aquatics culture facility and acclimated in a separate tank for one week prior to exposure. Exposures were carried out in two liter beakers containing one liter of the following treatments: embryo rearing medium (ERM: 17.1 mM NaCl, 402 μM KCl, 272 μM $CaCl_2 \cdot 2H_2O$, 661 μM $MgSO_4 \cdot 7H_2O$, pH 7.2), Raleigh, NC Reuse Water (RUW), 1 ppb clofibrate in RUW and 10 ppb clofibrate in RUW; each beaker contained 0.001% DMSO after dosing. Nine fish were randomly selected and added to each beaker, with three replicate beakers per treatment. Fish were maintained at room temperature on a 16h:8h light:dark cycle. To ensure water quality, 80-90% of the exposure solutions were replaced daily. Fish were fed approximately 50 mg Otohime B1 aquarium feed (Reed Mariculture) per beaker on day one and day three of exposure. Three fish per beaker were sacrificed on days one and three for RNA isolation. For RNA isolation, fish were anaesthetized in ice-cold ERM and livers excised and immediately frozen in liquid nitrogen and stored in cryovials at -80°C for one week prior to RNA isolation.

Table 4-1. Dose Exposure Matrix for Reuse Water Microarray Analysis.

Time	Raleigh Reuse/ DMSO Control	Raleigh Reuse/Clofibrate 0.001	Raleigh Reuse/Clofibrate 0.01	Laboratory rearing media DMSO Control
24hrs	3 Biological reps @ 3 fish each (9 fish) 2 technical array reps/biological rep (6 arrays)	3 Biological reps @ 3 fish each (9 fish) 2 technical array reps/biological rep (6 arrays)	3 Biological reps @ 3 fish each (9 fish) 2 technical array reps/biological rep (6 arrays)	3 Biological reps @ 3 fish each (9 fish) 2 technical array reps/biological rep (6 arrays)
72hr	3 Biological reps @ 3 fish each (9 fish) 2 technical array reps/biological rep (6 arrays)	3 Biological reps @ 3 fish each (9 fish) 2 technical array reps/biological rep (6 arrays)	3 Biological reps @ 3 fish each (9 fish) 2 technical array reps/biological rep (6 arrays)	3 Biological reps @ 3 fish each (9 fish) 2 technical array reps/biological rep (6 arrays)

4.2.2 RNA Isolation

Pooled livers were homogenized in 1 mL RNABee (TelTest) using a Polytron stainless steel homogenizer cleaned sequentially with RNaseZAP (Sigma), DEPC-treated water and sterile de-ionized water. 200 μL chloroform was added to each sample and the samples were stored on ice for 5 minutes. Samples were centrifuged at 4°C and the aqueous phase was added to 500μL isopropanol. Samples were centrifuged at room temperature and the RNA pellet was resuspended in 1 mL 75% ethanol. Samples were centrifuged at 4°C and pellet was dried then resuspended in RNase free water. Following this step, RNA isolation was carried out following manufacturer's instructions using an RNeasy Mini Kit (Qiagen). Prior to sample elution, contaminant DNA was removed by performing an on-column digestion with an RNase-free DNase Set according to manufacturer protocol (Qiagen). Samples were eluted in 52°C RNase-free water. RNA quantity and 260/280 ratios were determined using a NanoDrop® ND-1000 spectrophotometer. Total RNA was stored at -80°C prior to cDNA array hybridizations.

4.2.3 cDNA Arrays

186-gene membrane microarrays were constructed using subtraction libraries generated previously (see Chapters 2.0 and 3.0). To construct cDNA membrane arrays, stock colonies from all subtraction libraries were individually transferred with sterile toothpicks to two 96-well microplates containing 100 μL LB media with ampicillin (100 μg/mL) per well and regrown overnight at 37°C with constant agitation at 150 rpm. Following overnight growth, all new stock colonies were stored in 60% glycerol at -80°C until PCR amplification. All cDNA clones were simultaneously PCR-amplified in duplicate using two 96-well PCR plates. For each 50 μL PCR reaction, cDNA clones were PCR amplified using 4 μL bacterial growth, 5 μL 10X Advantage 2

Buffer, 1.5 µL 10µM NP-1 primer, 1.5 µL 10µM NP-2R primer, 1µL 10µM dNTPs, 36.5 µL sterile water, and 0.5 µL 50X Advantage 2 Polymerase Mix (BD Biosciences). PCR reaction conditions were: 94°C for 30s followed by 35 cycles of of 95°C for 10s and 68°C for 3 min. PCR products were checked for purity by resolving nine randomly selected wells from each PCR plate (18 total PCR products) on a 1.2% agarose gel. Once quality was established, PCR amplified cDNAs were individually purified using a QIAquick PCR purification kit (Qiagen) and quantified in 1 µL aliquots using a Nano Drop ND-1000 UV/Vis spectrophotometer (NanoDrop Technologies). For each cDNA membrane array, 25 ng of purified cDNA was denatured at 65°C for 10 min. with 0.6M NaOH and chilled on ice for 2-3 min. Duplicate sets of probe were spotted onto positively charged nylon membranes (75 mm X 115 mm) (Roche Diagnostics) using a sterile 96 pin Slot Pin Multi-Blot™ Replicator (V&P Scientific) with a 2 µL delivery volume. To assess intra- and inter-array variability, array controls were spotted (25 ng/spot) on each membrane array. Array controls included: Arabidopsis Cab1 cDNA (Accession ID: X56062) (6 spots/array) (Spot Report-3 Array Validation System, Stratagene) and salmon sperm DNA (28 spots/array) (Sigma-Aldrich). Spotted membranes were neutralized in 0.5 M Tris/1.5 M NaCl for 30s, rinsed in water for 2 min., and cDNAs UV cross-linked to membranes prior to drying and storage. As two arrays per replicate RNA pool were used for analysis, a total of 48 identical membrane arrays (6 arrays per dose-time condition) were spotted and stored in a dry, sealed environment at room temperature until hybridizations.

^{32}P-labeled cDNA probes for array hybridizations were generated in a labeling reaction containing array control mRNA (Arabidopsis Cab1 mRNA) (Stratgene) and total sample RNA (ERM, RUW, or Clofibrate/RUW treated liver RNA); Arabidopsis Cab1 mRNA was used as an internal standard to control for cDNA labeling and hybridization efficiency. For reverse transcriptase (RT) labeling reactions, 2 µg total RNA, 1 ng Arabidopsis Cab1 mRNA, and 4µL 10 µM oligo(dT) primer were mixed and denatured at 65°C for 5 min. After cooling on ice for 5 min., 8µL 5X First Strand buffer (Invitrogen), 4µL 0.1 M DTT, 2µL RNaseOUT™ Ribonuclease Inhibitor (40 U/µL) (Invitrogen), 2µL modified dNTPs (10mM dGTP, 10mM dCTP, 10mM dTTP, and 0.05 mM dATP), 40µCi [^{32}P]dATP (Amersham Biosciences), and 1 µL M-MLV RT (200 U/µL) (Invitrogen) were mixed thoroughly with the above, and the final mixture incubated at 42°C for 1.5h. Following incubation, all ^{32}P-labeled cDNAs were purified using G-25 columns (Amersham Biosciences), and probe radioactivity quantified using a bench-top Beckman Coulter LS 6500 scintillation counter (Beckman Coulter). Using glass hybridization tubes, membrane arrays were individually hybridized with 5 mL hybridization buffer (ExpressHyb Hybridization Solution, BD Biosciences) and 100µL blocking solution (10X SSC/5.5 mg/mL denatured salmon sperm DNA) at 72°C while rotating for 1.5h. Based on ^{32}P-specific radioactive counts for each triplicate set of labeled probes, equal quantities of ^{32}P-labeled cDNA probe from each treatment replicate labeling reaction diluted in 100µL blocking solution were hybridized to appropriate membrane arrays overnight while rotating at 72°C. For each RNA sample, two replicate membranes containing duplicate spots per membrane were hybridized with ^{32}P-labeled cDNAs in separate hybridization tubes and ovens to control for random technical errors. As a result, six membranes were hybridized for each dose-time condition, totaling 48 membrane hybridizations for the entire experiment and 12 replicate spots per dose-time condition for each gene. Following hybridization, membrane arrays were washed four times with 2X SSC/0.5% SDS for 30 min./wash and two times with 0.2X SSC/0.5% SDS for 30 min./wash. Following washes, hybridized membranes were plastic-wrapped, exposed onto phosphor screens for 72h and scanned using a Storm 860 Phosphoimager (Molecular Dynamics).

4.2.4 Statistics

Raw spot density data for each membrane were individually quantified by ArrayGauge v2.1 (FugiFilm), exported as a spreadsheet into Microsoft Excel, and coded by gene, treatment, treatment replicate, day, membrane replicate and spot replicate. For each membrane, spot intensity values were normalized by dividing by mean intensity values of Arabidopsis Cab1 on the same membrane. This adjusted the raw intensity values from each membrane for differences in cDNA labeling and hybridization among membranes. All Cab1-normalized values were $\log_2$ transformed to ensure model assumptions of linearity and equal variance. Log_2-transformed data were clustered by gene using un-centered correlation in with average linkage clustering in Eisen's Cluster and TreeView (Eisen et al., 1998). To test for statistical significance, $\log_2$-transformed data were imported into SAS v9.1 (SAS Institute, Inc.) and tested for main effects of day or treatment and day-treatment interaction effects using a mixed linear model (α=0.05) (a MIXED procedure in SAS) (Wolfinger et al., 2001). This model considers both fixed and random effects parameters associated with known a priori assigned variables (gene, day and treatment) and unknown random variables (treatment replicate, membrane replicate, and spot replicate), respectively, assumed to impact the variability of the data; thus, since these data were generated using a nested experimental design, we used a mixed linear model to account for variability linked with each level of hierarchy. Bonferroni-adjusted multiple comparisons of least square means were used to account for high false-positive rates associated with post-hoc pair-wise comparisons. Mixed linear models and multiple comparisons were applied at the level of gene expression pattern and individual gene response to test significance at different organizational hierarchies in the data. To evaluate entire gene expression patterns across dose and time, principal component analysis (PCA) was performed using PRINCOMP procedure (α=0.05) in SAS, and was based on time-matched treatment to control ratios calculated from mean normalized data from each gene.

4.3 Results and Discussion

Animal survival for the entire 3-d reuse water exposure period was 100% across all treatments. Water parameters determined for ammonia, alkalinity, hardness, TOC, turbidity, pH, conductivity, and specific ion concentrations were within acceptable tolerance for exposure studies. Approximately 80% of water was renewed daily to ensure minimal effect(s) due to increase in nitrogen accumulation. Additionally no sublethal or behavioral effects were observed, fish continued feeding, maintained buoyancy, normal swimming and tracking, and avoidance response. These results suggest that reuse water quality was sufficient for the designed exposure period.

In this study, the researchers used custom hepatic 186-gene microarray to assess fibrate-induced transcriptional changes on post-exposure days 1 and 3. This study demonstrated that acute aqueous exposure resulted in significant early transcriptional events in medaka liver. Based on the genes surveyed, the most pronounced changes in gene expression occurred on day 1 and were associated with 27-28 different biological processes in the liver (Figure 4-1). These data suggest that clofibrate affects a wide range of cellular mechanisms by 24 h post-exposure. Following multiple comparisons of microarray patterns across dose-time groups, overall hepatic transcriptional changes were significantly different across each microarray treatment when comparing each dose to control (single cross) and adjacent treatment (double cross) within each day. Based on microarray-wide normalized spot intensity data, there were significant main effects of dose and day and significant dose-day interaction effects when comparing entire gene expression profiles (Figure 4-1). These data suggest that experimental design that incorporates both biological and technical replication results in reduced variability and high statistical significance. Interestingly, our array comparisons additionally demonstrate that reuse water alone produces statistically different gene expression changes when compared to control laboratory waters on day three (Figures 4-1 and 4-2A). These results suggest that components of urban waste may be present and contribute to observed changes in gene expression. As such contents of reuse water alone are capable of eliciting a gene response.

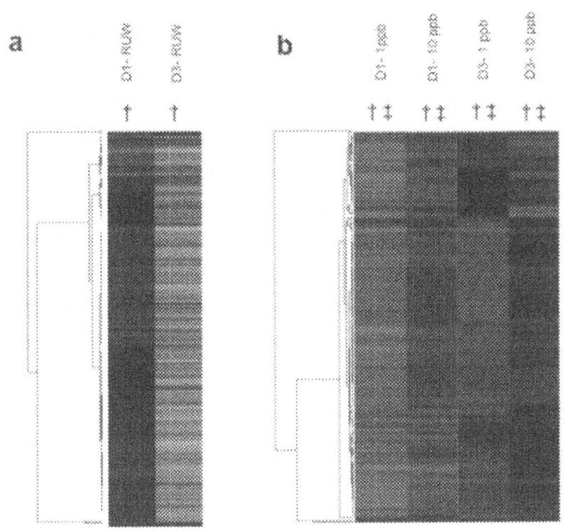

Figure 4-1. Gene Expression Patterns in RUW and RUW/Clofibrate Treated Medaka Livers.
a) RUW treatments: hierarchical cluster of expression data by gene based on mean fold changes relative to ERM-treated controls. Red denotes up-regulation of gene and green denotes down-regulation. Single cross (†) indicates that the treatment array mean is significantly different from time-matched ERM control array mean. b) Clofibrate/RUW treatments: hierarchical cluster of expression data by gene based on mean fold changes relative to RUW-treated controls; RUW treatment was chosen as the control because the clofibrate doses were applied in RUW. Red denotes up-regulation of gene and green denotes down-regulation. Single cross (†) indicates that the treatment array mean is significantly different from time-matched RUW control array mean. Double cross (‡) indicates that the treatment is significantly different from the other time-matched dose array mean.

Gene expression changes were most pronounced on day 1, low dose clofibrate treated fish. Within day-1, there was a dose response in overall gene expression data where hepatic gene expression patterns from 1ppb and 10.ppb treatments were both significantly different from control reuse water. Through statistical filtering of gene microarray data, there were 116 and 45 significant transcript responses ($\alpha = 0.05$) on day 1 in fish treated with 1 and 10.0 ppb respectively (Figure 4-2B). These significant responses were represented by most ontological categories demonstrating that treatment affected a wide range of cellular processes by 24 h in the liver. Gene expression changes in all treatments on day 3 were significantly decreased relative to day 1. By day 3, there was a sharp decrease to 6 and 1 significant transcriptional responses in fish treated with 1, and 10.0 ppb respectively (Figure 4-2B). At the high dose on day 3 gene expression patterns approached controls values in fish exposed with only one significant transcript.

Figure 4-2. Number of Genes Responding at Significance Levels of p=0.1, p=0.05 and p=0.01. a) RUW treatments compared to ERM controls b) Clofibrate treatments compared to RUW.

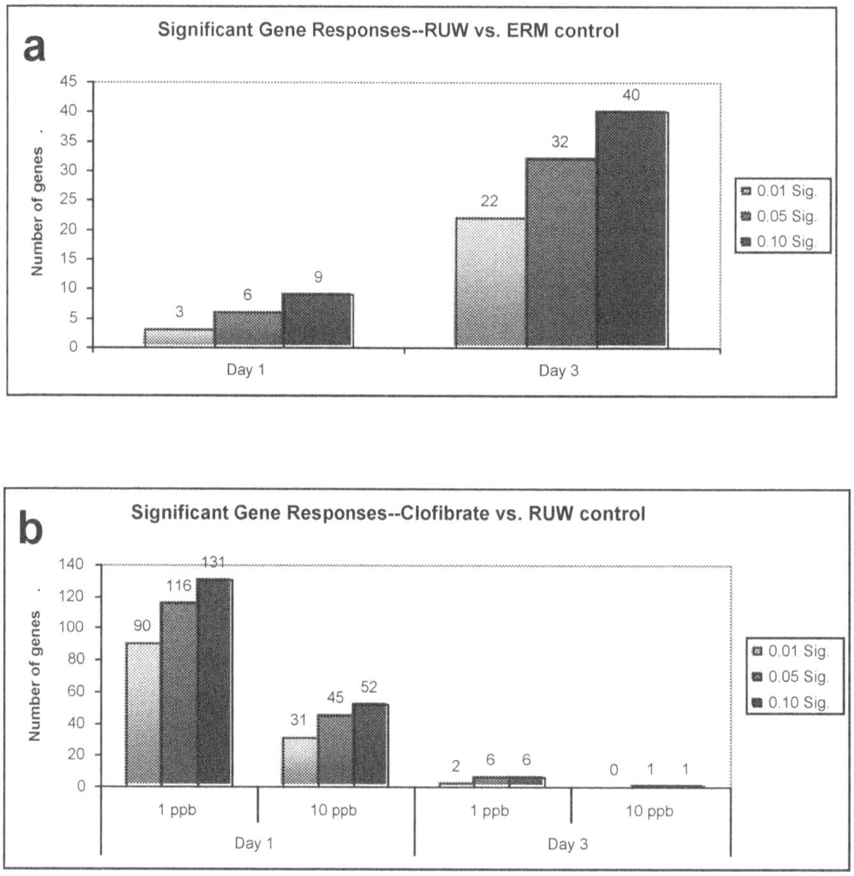

Unsupervised data analysis using both hierarchical clustering and principal component analysis were use in the absence of *a priori* information regarding treatment or gene annotation. The goal of gene expression data clustering is to group together genes and or treatments that exhibit similar expression patterns. It is useful for discovering the behavior of an expression matrix and aids in reducing the dimensionality of the data set were all genes on an array are represented by a cluster of genes that behave similarly.

Clustering in Figure 4-3 is based upon a correlation method that employs an agglomerative process that successively groups similar objects in a pair-wise iterative fashion until all data is included. The dendrogram is organized with treatments, both day and dose, across the top of the figure and genes along the left hand side. Branch lengths of the tree represent degrees of similarity between adjacent treatments or gene expression patterns. Hierarchical clustering was applied to gene expression profiles across fibrate treatments as a ratio of clofibrate treatment (in reuse water)/reuse water controls. All genes represented in the cluster exhibited statistically significant expression alterations in at least one condition relative to the control.

Clustering along the X axis demonstrates a distinctive relationship between gene expression and time and dose of treatment. The resulting hierarchical ordering of the cluster tree suggests that the maximal correlation is observed between doses on day one. Next, day 1 low and high dose are most similar to day 3 low dose followed by an independent node for day 3 high dose. This is consistent with observed ratio values were day one exhibits a higher

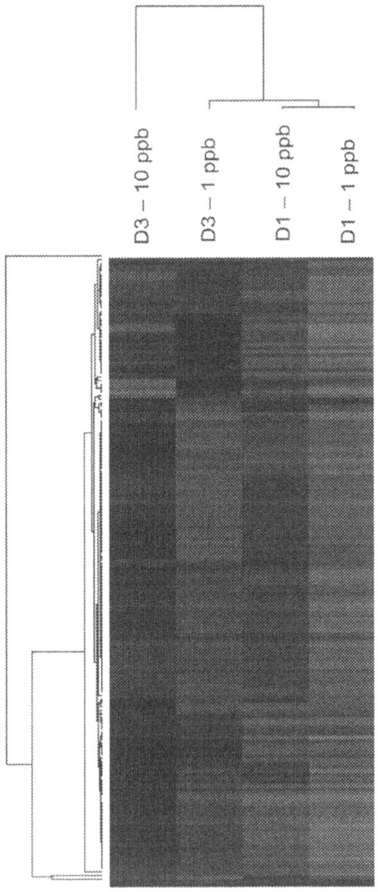

Figure 4-3. Gene Expression Patterns in Clofibrate Treated/RUW Control Treated Medaka Livers.
Hierarchical cluster of expression data by gene and array based on mean fold changes relative to RUW-treated controls; RUW treatment was chosen as the control because the clofibrate doses were applied in RUW. Red denotes up-regulation of gene and green denotes down-regulation.

frequency of up regulated genes as compared to day 3. Gene expression ratios decrease over time with a significant number of genes down-regulated by day three high dose. These gene expression patterns reveal several key events following treatment including time dependent up-regulation, time dependent down regulation and transient up or down regulation. Similar data trends were evident from PCA where the first principal component accounted for 37% of the data variability and was largely influenced by the strong temporal day-1 response in all both treatments (Fig. 4-4). After considering the day-1 contribution to the variation in gene expression data, the second principal component was influenced by dose for the day 1 treatment accounting for 19% of the total data variability. Day 3 however did not demonstrate any dose response relationships, similar to that observed in hierarchical clustering. Based on both unsupervised methodologies the predominant parameter driving organization of the data set is time. However within the hieratical clustering and PCA a clear dose response relationship is observed. What is particularly interesting is the gradual decrease in up regulated genes across time and dose. In Figure 4-3 it is apparent that day one low dose results in a large and significant up-regulation of in gene response. This is followed by a pronounced decrease in the high dose, followed by further decrease in day 3 low dose. The degree of up regulation appears to be driving clustering arrangement. These results however, are counterintuitive as it would normally be expected to see an increase in up regulation with both increasing dose and time of exposure. Studies conducted in laboratory water exhibit a similar pattern of hierarchical organization with time to response the predominant factor driving gene expression in this study. Previously it was reasoned that this was due to high dose ranges, saturating the gene expression response. However, in this and other studies the researchers observed distinct temporal effects on gene expression. This most like reflects the temporal relationship between xenobiotic exposure and gene expression where initial gene expression changes are followed by a series of secondary changes resulting in temporal similarities that supercede dose. The fact that in this study a dose response is observed at lower concentrations suggests that concentrations used in the initial study may have been too high. This is further supported by a recent study in the researchers' laboratory where both temporal and dose-related responses were observed following a single low dose exposure to TCDD (Volz et al., 2006). This suggests that a lower concentration series may in fact produce a more significant distinction between dose and time. While in general, clear relationships to dose response are favorable in this type of study, these results suggest that a molecular response is highly sensitive and that microarray analysis will provide a significant means of determining gene expression changes at lower concentrations as demonstrated in this study at the 1 ppb dose. The route of exposure additionally needs to be considered as dosing was conducted as static renewal and thus organisms were continuously exposed. Constant exposure to lipophilic pharmaceuticals may result in altered pharmacokinetic parameters including absorption, distribution, metabolism and excretion (ADME). Illustrating this point is the observed decrease in some metabolic genes including phase I and phase two enzymes. Additionally genes associated with lipid metabolism were significantly altered in both doses on day one.

By comparison results in reuse water were significantly different than expected from laboratory water exposures. This discrepancy is most likely due to the presences of additional contaminants in the reuse water capable of eliciting a gene expression response. Naturally, when dealing with contaminant mixtures, a more complicated gene expression profile will be apparent. Examination of general trends demonstrates that in reuse water low and high dose day one exposures produce a highly up-regulated array profile. In comparison, laboratory water exposures result general downward trend in gene expression. With time (by day 3) however expression profiles for 10 ppb reuse water appears similar to that observed under laboratory conditions. Interpretation of results on an array basis are difficult, however, it was speculated that the constituents of the reuse water are obscuring fibrate specific gene expression data particularly at the lower doses and later time points. In fact analysis of arrays conducted with fish exposed solely to reuse water demonstrate a high degree of up regulation on day 1 similar to that observed when 1 and 10 ppb of clofibrate are added however, overall response is less intense and not statistically significant. In longer dosing periods i.e. day 3, the degree of up-regulation subsides and profiles appear more similar to spiked laboratory water.

On a gene by gene basis, the researchers initially identified several candidate markers of fibrate exposure in laboratory water (see chapter three). While these markers were consistent across all laboratory exposures, comparison to different classes of potential aqueous contaminants such as estrogens suggested there maybe some overlapping gene responses. In response, principal component analysis was conducted to ensure that the two pharmaceutical classes formed distinct clusters. Following this analysis data sets were filtered for genes with overlapping expression values for each class. As a result a set of potential gene markers were chosen that do not display any overlap between fibrates and estrogens. One caveat to this data set however, is that only two classes of compounds were initially tested. Thus presence of additional contaminants in reuse water may influence gene expression of these marker sets skewing analysis and interpretation. When examining gene expression profiles in reuse water several observations were noted:

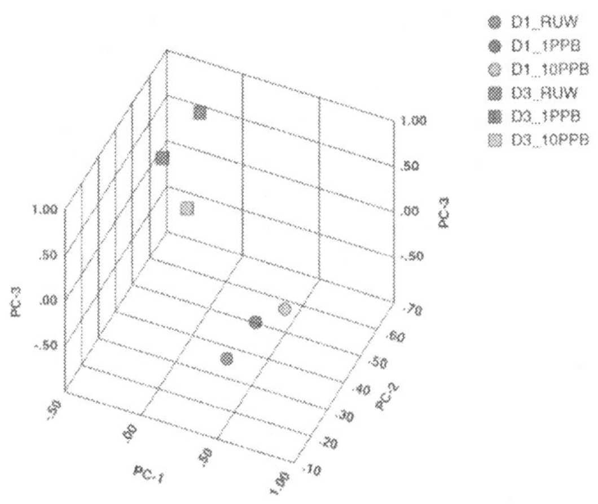

Figure 4-4. Principal Component Analysis. Clofibrate in reuse water control, 1 ppb, and 10ppb on days 1 and 3.

Comparisons of day one exposures of reuse water *without* addition of clofibrate yielded few statistically significant, differentially expressed genes when compared to laboratory water controls. This suggest that following a 24-hour exposure gene expression events are occurring but are not significantly different from controls. This is to be expected given potential stress related effects from handling and rearing conditions. Day one 1 ppb clofibrate in reuse water yielded numerous statistically significant genes differentially expressed compared to reuse water controls. Of these 69% of the genes represented previously identified "fibrate responsive" marker genes in the laboratory experiments. These results suggest that a clear and significant effect could be determined by analysis of our select genes within our established training set from laboratory exposures. Day one, 10ppb clofibrate in reuse water yielded several statistically significantly differences when compared to reuse water alone, of which 65% represented "fibrate responsive" marker genes. However overall the day one 10 ppb exposure resulted in fewer total statistically significant genes than observed in the lower 1.0 ppb dose. In previous studies the researchers have consistently observed an increase in numbers of statistically significant genes with increasing dose. The reason for the decrease in this study is unknown, but may be associated with either water matrix effects or complications of contaminant mixtures present in the reuse water.

4.4 Concluding Remarks

In this research study the researchers examined the utility of using a gene expression changes via a microarray platform for the assessment and detection of fibrate chemicals in laboratory and reuse water. Results from this study demonstrate that the microarray array procedure and analysis provides robust and statistically significant process for establishing compound specific gene expression changes. In Chapter 3.0, gene expression profiles were established for a set prototypic chemical contaminants found in some reuse water systems. In a limited data set these profiles were shown to be class specific. However, in some instances a few genes are coregulated by other suspected reuse water contaminants such as ethinyl estradiol. In Chapter 4.0 experiments were extended to assess gene expression changes in reuse water.

Analysis of reuse water spiked with cholfibrate suggests that gene regulation occurred in both an up and down direction. Gene expression changes were most pronounced on day 1 in the low dose treatment and in some instances a dose response effect was observed were hepatic gene expression patterns from 1ppb and 10.ppb treatments were both significantly different from control reuse water. Fewer statistical genes were observed on day three at both the high and low dose. This is like due to a high background of gene expression changes in the control exposures. Interestingly, when compared to laboratory water, the reuse water alone resulted in large number of statistically significant gene changes. The genes included several candidate markers for fibrates in addition to genes previously identified as responsive to estrogens and aromatic hydrocarbons (other studies in our laboratory), suggesting that reuse water used for this study contained a mixture of urban contaminants at concentrations capable of eliciting gene expression changes in our model system. The lack of statistically significant gene changes in day 3 may be a result of the high background activity of the reuse water alone. Thus it is likely that shorter exposure times will yield more significant results. However, in the most likely application of this technology, reuse waters with unknown contaminants will be compared to laboratory waters where background expression values will not interferer with analysis.

CHAPTER 5.0

RECOMMENDATIONS FOR FUTURE RESEARCH

Toxicogenomic projects that utilize differential gene expression approaches will evolve to address key components for gene discovery, pathway analysis, and biomarker determination in applied aquatic biology. The completion of genome sequences of a wide range of organisms is presenting new and exciting challenges to research in numerous disciplines including aquatic biology and environmental engineering. In the search to understand how complex organisms interact with the environment, it is necessary to understand the function of individual genes and proteins, cells and tissues, organs and organ systems, and further along higher levels of biological organization and complexity. The integration of several scientific disciplines with new paradigms in science including genomics, proteomics and metabolomics are become integral components of most environmental assessment and monitoring programs. With these, environmental scientists are well poised to establish a thorough knowledge of the complex interactions between environmental exposures, environmental engineering and environmental impacts. Ultimately, these approaches will provide highly sensitive and beneficial methodologies for assessment of environmental and human risk for waste, reuse and drinking water monitoring programs.

In the most likely scenario reuse waters with unknown contaminants will be compared to laboratory waters where background expression values will not interferer with analysis. With further development of microarrays and screening of prototypic contaminant classes, training sets can be established to further develop class prediction models for the identification of compounds- or compound classes from complex mixtures of water contaminants. A key to this approach will be the ability to clearly identify non-overlapping gene markers that comprise an established gene expression profiles. This approach will enable a focus on the patterns of gene expression or expression profiles and observe distinct differences in gene regulation. Additionally, if gene annotations are examined, it becomes obvious that specific chemical classes will elicit defined metabolic pathways based upon their mechanism of action. As described in Chapter 2.0, fibrate drugs are PPARα agonist and thus result in expression of a defined set of genes and pathways associated with lipid metabolism and transport, amino acid metabolism and others that can be exploited for profile analysis.

In an effort to make this technology feasible for of water quality assessment the following research needs should be addressed:

- Identify suspect reuse water contaminants by examination of urban population pharmaceutical use patterns and demographics.
- Develop gene expression training sets by screening of prototypic contaminant classes, in species of interest.
- Ascertain dose response effects.

- Ascertain time to response effects
- Clearly determine non-overlapping gene markers that comprise an established gene expression profile.
- Validate gene expression events by analysis of biochemical pathways based upon known mechanisms of pharmaceutical action.
- Determine effects of reuse water matrix on gene expression (TOC, alkalinity, DO etc.)
- Establish a smaller set of gene expression changes that are indicative of contaminant exposure. Once established, conduct real time polymerase chain reaction (PCR) in place of microarray studies. This analysis will be better suited to the technical capabilities of interested utilities.

APPENDICES

Appendix A-1

Suppressive subtractive hybridization identified responsive genes from adult male medaka liver. Freq = frequency in the subtracted and sequenced library. Italicized genes are those shared by two organs and bolded genes are those shared by all three organs.

Gene Name	E value	Freq	Gene Ontology™
Acetyl-CoA carboxylase	2.00E-70	1	Fatty acid/lipid metabolism
Aldolase-B	3.00E-41	1	Glycolysis
Alpha-1-antiproteinase-like protein	1.00E-57	1	Proteinase inhibitor
Apolipoprotein AI precursor	9.00E-31	4	Lipid transport
Apolipoprotein B	8.00E-54	2	Lipid transport
Beta-carotene dioxygenase 2	5.00E-22	1	Carotene/lipid metabolism
C1q-like adipose specific protein	3.00E-32	4	Complement activation
Complement	1.00E-101	1	Complement activation
Complement bf/C2	0.00E+00	1	Complement activation
Complement C1r/s-A isotype	7.00E-51	2	Complement activation
Complement C3-1	3.00E-90	12	Complement activation
Complement C3-1, C3-2	0.00E+00	1	Complement activation
Complement C3-2	1.00E-85	3	Complement activation
Complement C9	1.00E-15	1	Complement activation
Complement component C9	1.00E-66	1	Complement activation
Complement factor H-related protein	9.00E-16	5	Complement activation
Coregulation factor V	2.10E-02	1	Unkown
Cytochrome P450 3A40	5.00E-73	1	Drug/toxicant metabolism
Cytochrome C oxidase subunit II	1.00E-15	1	Ion/electron transport
Ependymin	9.00E-17	3	Cell matrix adhesion
Hypothetical 18K protein (goldfish mitochondria)	1.00E-17	1	Unkown
Insulin-like growth factor binding protein 2	3.00E-16	1	Cell growth regulation
Inter-alpha trypsin inhibitor (heavy chain 3)	9.00E-41	1	Protease inhibitor
Liver basic fatty acid binding protein	2.00E-51	1	Lipid transport
Methionine adenosyltransferase II (alpha)	1.00E-28	1	One-carbon cmpd. metabolism
MHC class 1	7.00E-69	2	Immune response
Pentraxin	2.00E-38	2	CNS development
Phosphoenolpyruvate carboxykinase (cytosolic)	2.00E-15	1	Lipid metabolism
Precerebellin-like protein	3.00E-32	3	Immune response
Serotransferrin precursor	5.00E-99	15	Iron transport
Transferrin	4.00E-98	4	Iron transport
Transposase (von Hippel-Lindau binding protein)	9.00E-17	1	Transport/chaperone
UDP glucuronosyltransferase 1A10	1.00E-38	1	Drug/toxicant metabolism
Vitronectin	1.00E-11	1	Immune response
Wap65 water temp acclimation related protein	6.00E-83	3	Immune response

Appendix A-2

Suppressive subtractive hybridization identified responsive genes from adult male medaka brain. Freq = frequency in the subtracted and sequenced library. Italicized genes are those shared by two organs and bolded genes are those shared by all three organs.

Gene Name	E value	Freq	Gene Ontology™
Aldolase A	2.00E-45	1	Glycolysis
Alpha tubulin	1.00E-163	1	Microtubule-based movement
Ancient ubiquitous protein 1 isoform 1	7.00E-19	1	Unkown
ATP dependent helicase (RNA helicase)	5.00E-91	1	DNA binding/transcription reg.
CED-12 (EMOL)	5.00E-49	1	Apoptosis/phagotosis
Cytochrome b	1.00E-176	1	Ion/electron transport
Cytochrome c oxidase subunit I	1.00E-65	1	Ion/electron transport
Dynactin subunit p25	4.00E-12	1	Microtubule-based movement
Enolase beta	1.00E-37	1	Glycolysis
G(3)DPH	1.00E-145	1	Glycolysis
Glutamine synthetase	3.00E-81	1	Nitrogen metabolism
GSH-T	0.011	1	Drug/toxicant metabolism
Hypothetical protein FLJ10979	7.00E-39	1	Unkown
MHC class 1	2.00E-91	2	Immune response
Mitochondrial DNA	0.00E+00	1	Unkown
Modulator of E2 induced transcription	3.00E-04	1	Transcription regulation
Muscle-type creatine kinase	0.00E+00	1	Phospohcreatine metabolism
NADH dehydrogenase 1 alpha	3.00E-11	1	Ion/electron transport
Phosphohippolin	2.00E-20	1	Ion/electron transport
Phospholipid scramblase Zic1	1.00E-149	1	Membrane organization
Protease (macropain) 26S ATPase	6.00E-05	1	Protease/protein modification
Ribonuclease P protein 3	2.00E-18	1	tRNA processing
Sarcoplasmic reticulum calcium ATPase	2.00E-80	1	Ion/electron transport
Serine \threonine phosphtase (PP2A)	2.00E-26	1	Signal transduction
Sodium/postassium ATPase alpha	1.00E-111	1	Ion/electron transport
Thyrotropin beta chain precursor	4.00E-40	1	Signal transduction
Tumor necrosis factor (endothial)	9.00E-18	1	Immune response
Type I inositol 5-phosphatase	4.00E-19	1	Cell comm../ Signal transduct.
Ubiquitin-conjugating enzyme E2	1.00E-83	1	Protein catabolism

Appendix A-3

Suppressive subtractive hybridization identified responsive genes from adult male medaka testis. Freq = frequency in the subtracted and sequenced library. Italicized genes are those shared by two organs and bolded genes are those shared by all three organs.

Gene Name	E value	Freq	Gene Ontology™
2,3-Bisphosphoglycerate synthase	2.00E-80	1	Glycolysis
2610207P08Rik protein	1.00E-21	1	Unknown
26S Proteasome non-ATPase regulatory subunit	1.00E-67	1	Protein catabolism
Arginine methyltransferase 1b	9.00E-31	1	Catalytic transferase activity
Beta 1,4-galactosyltransferase	8.00E-18	1	Metabolism/2° sex char. devel.
BTG associated nuclear protein	5.30E-02	1	Transcription regulation
Carbonic anhydrase IV precursor	1.00E-11	1	Complement activation
Cathepsin C	3.00E-19	1	Proteolysis/protein modification
CED12 A PH domain	3.00E-09	1	Complement activation
Chloride intracellular channel 4	1.00E-40	1	Ion/electron transport
Chromobox protein	5.00E-18	1	Transcription regulation
Cofactor reg. SP1 transcript activation sub 9	2.00E-38	1	Transcription regulation
Complement C3	0.00E+00	1	Complement activation
Cytochrome c oxidase (COX17)	2.00E-24	1	Ion/electron transport
Fizzy related protein (FZR1)	8.00E-20	1	Cell cycle regulation/apoptosis
G(3)DPH	1.00E-122	1	Glycolysis
GPI-anchored membrane protein 1	1.00E-27	1	Apoptosis/neg. cell cycle reg.
Guanine nucleotide binding protein	2.00E-22	1	Signal transduction
Hypothetical protein (human)	3.00E-58	1	Unknown
Inosine monophosphate dehydrogenase (IMP)	1.00E-40	1	Protein metabol./modification
Makorin 1	1.00E-13	1	Protein metabol./modification
Male sex determining protein (DMRT1Y)	2.00E-87	3	Sexual differentiation
Membrane translocase (34kD)	3.00E-42	1	Protein folding./modification
MHC class 1	1.00E-159	2	Immune response
Mitochondrial DNA	4.90E-02	1	Unkown
N-acetylglucosaminyltransferase	2.00E-63	1	Protein glycosyl../modification
Organic cation transporter	3.00E-52	1	Ion/electron transport
Parvalbumin	8.00E-47	1	Muscle development
PMP70 ATP binding cassette	4.00E-79	1	Peroxisome transport/org.
Predicted protein DR/R	7.00E-19	1	Unkown
Predicted protein RINKN	1.00E-21	1	Unkown
Protease (macropain) 26S ATPase	4.00E-03	1	Protease/protein modification
Protein phosphatase 1 regulatory (inhibitor)	5.00E-27	1	Signal transduction
Prothymosin alpha	1.00E-02	2	Cell cycle/transcription reg.
Rho GTPase activating protein	2.00E-04	1	Unknown
Retinoid X receptor (RXR)	1.00E-20	1	Transcription regulation
Seminal plasma glycoprotein	8.00E-52	1	Unknown
Similar to tubulin, alpha 1	2.00E-19	1	Microtubule-based movement
Sodium/postassium ATPase alpha	1.00E-119	1	Ion/electron transport
Split hand and fot disease BCRA2 DSS1	4.00E-37	1	Unknown
Translation elongation factor eEF-1	6.00E-12	1	Translat./protein modification
Translation initiation factor IeF4E	5.00E-33	1	Translat./protein modification

Appendix A-4

Cluster analysis of gene function. Hierarchical Clustering correlation-average linkage of Ciprofibrate day one 0.01-1.0 ppm

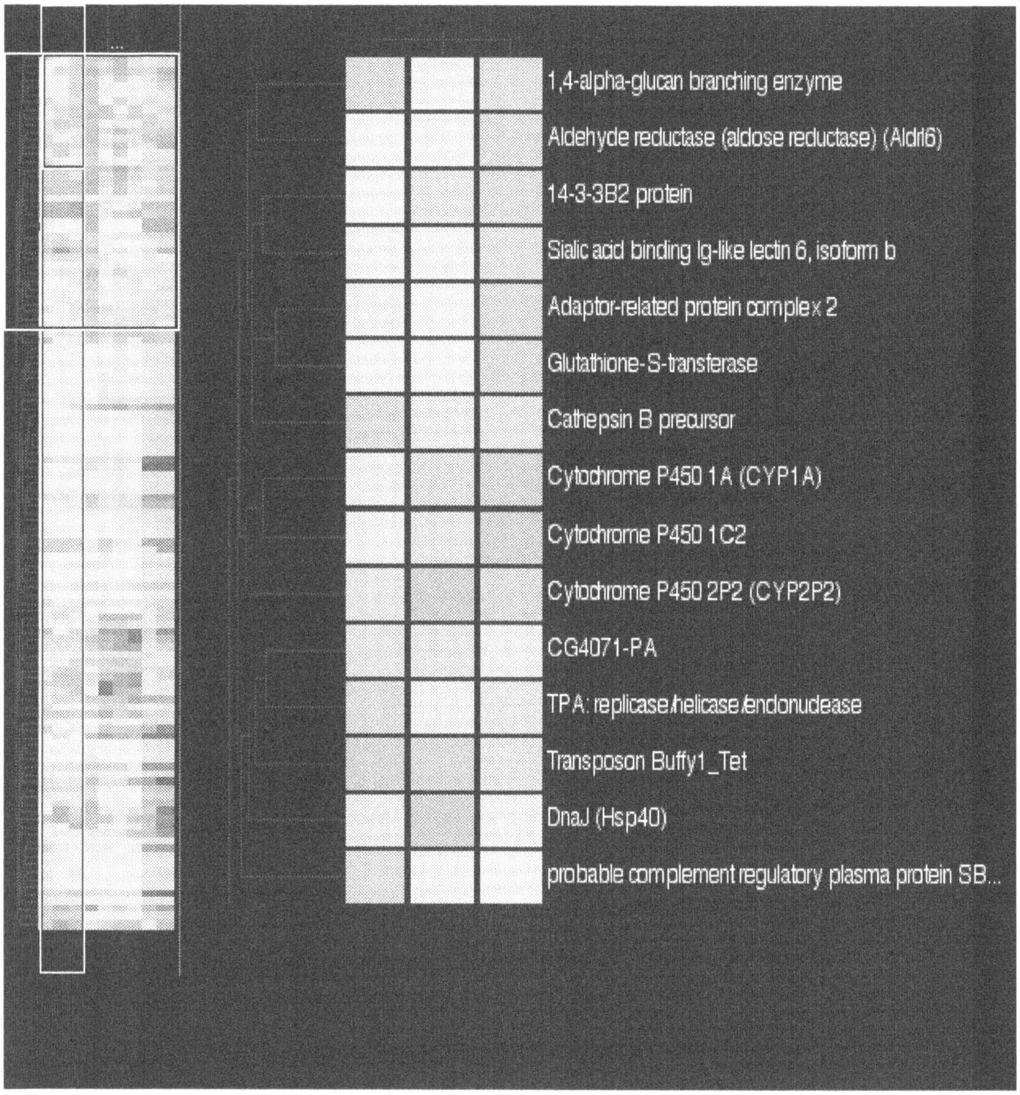

Figure A-4.

Appendix A-5

Gene Expression Profiles (up-regulated) Based on Common Gene response and Statistical Significance for Ciprofibrate and Clofibrate treated liver.

CF/CLO D1 0.01-1.0 UP	CF/CLO D3 0.01-1.0 UP	CF/CLO D7 0.01-1.0 UP
	1,4-alpha-glucan branching enzyme	
1-Cys Peroxiredoxin	1-Cys Peroxiredoxin	
acetyl-CoA carboxylase		
		acyl-Coenzyme A oxidase 3, pristanoyl
		alanine:glyoxylate aminotransferase
	Aldehyde dehydrogenase 12	
	Aldehyde reductase (aldose reductase) (Aldrl6)	
Aldolase A		
	aldolase-B	
	Annexin max3	
	Apolipoprotein E	
		Arrestin domain containing 2
	ATP binding cassette	
	Bile salt-activated lipase	
	Calmodulin	
	Calreticulin (CRT)	
	Cathepsin B precursor	
	Cathepsin C	
	CDC-like kinase 4 (Clk4) protein	
	Clone DKEY-202E22 in linkage group 21 [Danio rerio]	
	coagulation factor X precursor	
Coagulation factor XIII b subunit precursor		
	Complement C4	
		complement factor H-related protein
	Cone potassium-dependent sodium-calcium exchanger	
	Cytochrome P450 1A (CYP1A)	
	Cytochrome P450 1B (CYP1B)	
	Cytochrome P450 1C2	
		Cytochrome p450-cc24
	Dystroglycan 1	
		Eef2-prov protein
	Etr1	Etr1
		Ferritin
		Fibrinogen alpha-E subunit
Hepatocyte growth factor-like 1 (hgfl1)	Hepatocyte growth factor-like 1 (hgfl1)	
		High choriolytic enzyme 1 precursor
	HNF4	
Hypothetical protein LOC402870 (Danio rerio)	Hypothetical protein LOC402870 (Danio rerio)	
Hypothetical protein MGC63938 (Danio rerio)		
	Kallikrein	
	Leukocyte surface protein precursor	Leukocyte surface protein precursor
	Macrophage mannose receptor	

		Membrane guanylyl cyclase OIGC2
	Membrane-associated protein HEM-1	
	methionine adenosyltransferase II, alpha	
	MHC Class I	
	Mitogen-activated protein kinase phosphatase 2	
	Myo-inositol oxygenase (MIOX)	
Non-metastatic cells 4, protein		
		Nuclear receptor-binding factor 1
		P5-pending protein/protein disulfide isomerase-related protein
	parvalbumin	
	Parvalbumin (cyp c 1.01 gene)	
pentraxin [Cyprinus carpio]		pentraxin [Cyprinus carpio]
Phospholipid scramblase Zic1	Phospholipid scramblase Zic1	
Properdin factor, complement		
	PTPa and cGnRH-II	
	Putative disease resistance protein	
	Reticulon 4 receptor-like 2b	
		RIKEN cDNA 1600013K19
	RIKEN cDNA 2410046H15 (LOC306030)	
RXR, retinoid X receptor	RXR, retinoid X receptor	RXR, retinoid X receptor
	Sialic acid binding Ig-like lectin 6, isoform b	
Stromal cell derived factor 1		
	Thioredoxin domain containing 4 (endoplasmic reticulum)	
	THR	
	transposase/von Hippel-Lindau binding protein 1	
	Tyrosine aminotransferase	
UDP-glucoronosyl transferase	UDP-glucoronosyl transferase	
	Unknown (protein for MGC:55636) [Danio rerio]	Unknown (protein for MGC:55636) [Danio rerio]
	Wu:fa98f05	
Wu:fk63d09	Wu:fk63d09	

Appendix A-6

Gene Expression Profiles (down-regulated) Based on Common Gene response and Statistical Significance for Ciprofibrate and Clofibrate treated liver.

CF/CLO D1 0.01-1.0 DOWN	CF/CLO D3 0.01-1.0 DOWN	CF/CLO D7 0.01-1.0 DOWN
17-beta-hydroxysteroid dehydrogenase type 4		
Annexin max3		
Arrestin domain containing 2		
		ATP binding cassette -5
Bile salt-activated lipase		Bile salt-activated lipase -5
Cathepsin B precursor		
CDC-like kinase 4 (Clk4) protein		CDC-like kinase 4 (Clk4) protein -4
CG4071-PA		
Clone DKEY-202E22 in linkage group 21		
		Complement C4
Cone potassium-dependent sodium-calcium exchanger		
CXCR3 gene for CXC chemokine receptor		
Cytochrome P450 1A (CYP1A)		
Cytochrome P450 1B (CYP1B)		
Cytochrome P450 1C2		Cytochrome P450 1C2 -4
Cytochrome P450 2P2 (CYP2P2)	Cytochrome P450 2P2 (CYP2P2) (4)	Cytochrome P450 2P2 (CYP2P2) -4
DKFZP434K046 protein		
DnaJ (Hsp40)		
	Ependymin (4)	
Fetuin beta		
	FXR (6)	FXR -5
Glucose-6-phosphatase		
Glyceraldehyde 3-phosphate dehydrogenase		
		HNF4 -4
Inter-alpha trypsin inhibitor, heavy chain 3		
Kallikrein		
Leukocyte surface protein precursor		
	L-SF precursor (4)	L-SF precursor -4
LXR		
Membrane-associated protein HEM-1		
MHC Class I		
	Non-metastatic cells 4, protein (4)	
	organic cation transporter 4; (4)	
P5-pending protein/protein disulfide isomerase-related protein-parvalbumin		
		Parvalbumin (cyp c 1.01 gene) -4
	pentraxin [Cyprinus carpio] (5)	
PTPa and cGnRH-II		PTPa and cGnRH-II -4
Putative disease resistance protein		
Reticulon 4 receptor-like 2b		Reticulon 4 receptor-like 2b-4
RIKEN cDNA 2410046H15 (LOC306030)		RIKEN cDNA 2410046H15 (LOC306030)-4
	SERPINA10 (4)	

Syndecan binding protein (SDCBP)	
THR	
TPA: replicase/helicase/endonuclease	
Transcription factor nrf2	
Transposon Buffy1_Tet	
Tyrosine aminotransferase	Tyrosine aminotransferase-4
Unknown (protein for MGC:56244) [Danio rerio]	
VDR	
vitellogenin 1	
zt1.9 microsatellite sequence	zt1.9 microsatellite sequence

Appendix A-7

Gene Expression Profiles Based on Common Gene response and Statistical Significance for Ciprofibrate and Clofibrate treated Brain.

Gene Common up brain CF Clo D7 0.01-1.0 ppm	Gene Common down brain CF Clo D7 0.01-1.0 ppm
adenosine deaminase	Muscle-type creatine kinase (CKM1)
alpha-1-antiproteinase-like protein	glucose-6-phosphatase
ATP dependent helicase (RNA Helicase) DEAH	Tnfa-induced adipose-related protein
CED12 A PH domain	unnamed protein product
E25B protein	
extensin-like protein	
High mobility group protein-R	
hypothetical protein MGC5509 [Homo sapiens]	
Immunoglobin heavy chain	
MHC Class 1	
Mitogen-activated protein kinase phosphatase 2	
NADH dehydrogenase (ubiquinone) 1 alpha subcomp	
phosphohippolin	
Protease (prosome, macropain) 26S ATPase	
protease 1	
Protein 4.1	
PTPa and cGnRH-II	
ribonuclease, Rne/Rng family protein	
S100 calcium-binding protein	
Sarcoplasmic\endoplasmic reticulum Ca ATPase	
Sodium/Potassium ATPase alpha subunit	
techylectin-5A	
Translation initiation factor 4AIII	
Vacuole membrane protein 1	

Appendix A-8

Gene Expression Profiles Based on Common Gene response and Statistical Significance for Ciprofibrate and Clofibrate treated Testis.

Gene Common up CF/CLO D7 0.01-1.0ppm	Gene Common down CF/CLO D7 0.01-1.0ppm
Acyl-CoA dehydrogenase, medium-chain specific.	Aquaporin
Alpha-tubulin	centaurin, alpha 1 [Bos taurus]
amylase-3 protein [Tetraodon nigroviridis]	Danio rerio cDNA clone MGC:77094 IMAGE:6960468
Angiogenin	Delta 6-desaturase (FD6D)
BUB3 budding uninhibited by benzimidazoles 3-like protein (BUB3)	GPI-anchored membrane protein 1; GPI-anchored
Cell division kinase=cell cycle regulator cdc2 homolog	HMT1 hnRNP methyltransferase-like 2
C-myc binding protein	Keratin S8, type I
Compliment C3	KIAA1806 protein [Homo sapiens]
cytochrome b [Oryzias latipes]	polo-like kinase homolog; polo-like kinase homo
cytochrome P450 2P2 [Fundulus heterocli	serine/threonine kinase 19 isoform 1
Endothelin receptor	syntaxin 16A / syntaxin 16B)
erythrocyte membrane protein 1 (PfEMP1)	
fast skeletal muscle troponin I	
FLJ20433 protein	
Formyltetrahydrofolate dehydrogenase	
Guanylyl cyclase	
Hypothetical protein MGC56471 (Danio rerio)	
Hypothetical protein OB0705 [Oceanobacillus iheyensis HTE831]	
KIAA1228 protein	
MAD1 mitotic arrest deficient-like 1	
myosin light chain 2	
Na+/K+ transporting, alpha 1a.1 polypep	
N-acetylglucosaminyltransferase	
Nardilysin (N-arginine dibasic convertase) (NRD1 convertase)	
Parvalbumin	
Seminal plasma glycoprotein	
Split hand and fooot disease BCRA2 DSS1	
Sucrase-isomaltase, intestinal [Contains: S	
Sulfatase modifying factor 2	
Tektin 1	
Threonine/tyrosine-specific protein kinase	
Thymic dendritic cell-derived factor 1	
Translation initiation factor IeF4E	
Tudor repeat associator with PCTAIRE 2	
Unknown (protein for MGC:63961) [Danio rerio]	

REFERENCES

Anderson, M.J., H. Olson, F. Matsumura, and D.E. Hinton. 1996. In vivo modulation of 17-β-estradiol-induced vitellogenin synthesis and estrogen receptor in rainbow trout (*Oncorhynchus mykiss*) liver cells by B-naphthoflavone. Toxicol. Appl. Pharmicol. 137:210-218.

Ankely, G. 1998. Overview of a workshop on screening methods for detecting potential (anti-estrogenic/androgenic) chemicals in wildlife. Environ. Toxicol. Chem. 17:68-87.

Ara, I.A., H. Mitani, K. Naruse, and A. Shima. 1994. Relationship between the induction of proteins in the HSP70 family and thermosensitivity in two species of Oryzias (Pisces). Comp. Biochem. Physiol. B. Biochem. Mol. Biol. 109:647-654.

Arukwe, A., F. Knudsen, and A. Goksoyr. 1997. Fish zona-radiata (eggshell) protein: A sensitive biomarker for environmental estrogens. Environ. Health. Persp. 105:418-422.

Arukwe A., S.W. Kullman, and D.E. Hinton. 2001. Differential biomarker gene and protein expressions in nonylphenol and estradiol-17beta treated juvenile rainbow trout (Oncorhynchus mykiss). Comp. Biochem. Physiol. C. Toxicol. Pharmacol. 129:1-10.

Bartosiewicz, M., S. Penn, and A. Buckpitt. 2001. Applications of gene arrays in environmental toxicology: Fingerprints of gene regulation associated with cadmium chloride, benzo(a)pyrene, and trichloroethylene. Environ. Health. Persp. 109:71-74.

Bayne, C.J. and L. Gerwick. 2001. The acute phase response and innate immunity of fish. Dev. Comp. Immunol. 25:725-43.

Braumbeck, T.A., S. Teh, S.M. Lester, and D.E. Hinton. 1992. Ultrastructural alterations in hepatocyets of medaka (*Oryzias latipes*) exposed to diethylnitrosamine. Toxicol. Pathol. 20:179-196.

Brzostowski, M. 2001. On the Horizion. R&D. Directories. 7:30-38.

Cancio, I. and M.P. Cajaraville. 2000. Cell biology of peroxisomes and their characteristics in aquatic organisms. Int. Rev. Cytol. 99:201-93.

Carpy, S.A., W. Kobel, and J. Doe. 2000. Health risk of low-dose pesticides mixtures: a review of the 1985-1998 literature on combination toxicology and health risk assessment. J. Toxicol. Environ. Health. B. Crit. Rev. 3:1-25

Chomczynski, P. and N. Sacchi. 1987. Single-step method of RNA isolation by acid guanidinium thiocyanate-phenol-chloroform extraction. Anal. Biochem. 162:156-159.

Cooke, J.B. and D.E. Hinton. 1999. Promotion by 17-β-estradiol and beta-hexachlorocyclohexane of hepatocellular tumors in medaka, (*Oryzias latipes*). Aquatic. Toxicol. 45:127-145.

Crittenden, P.S., D. Reddy, W. Hand, and H. Arora. "Prediction of GAC Performance Using Rapid Small Scale Column Tests", Awwa Research Foundation, ISBN 0-89867-480-8, Denver CO. 1989.

De Benedetti, V.M., N. Biglia, P. Sismondi, and M. De Bortoli. 2000. DNA chips: the future of biomarkers. Int. J. Biol. Markers. 15:1-9.

De Koven, D.L., J.M. Nunez, S.M. Lester, D.E. Conklin, G.D. Marty, L.M. Parker, and D.E. Hinton. 1992. A purified diet for medaka (*Oryzias latipes*); refining a fish model for toxicological research. Lab. Anm. Sci. 42:180-189.

Diatchenko, L., Y.F.C. Lau, A.P. Campbell, A. Chenchik, F. Moqadam, B. Huang, S. Lukyanov, K. Lukyanov, N. Gurskaya, E.D. Sverdlov, and P.D. Siebert. 1996. Suppression subtractive hybridization: A method for generating differentially regulated or tissue-specific cDNA probes and libraries. Proc. Natl. Acad. Sci. USA. 93:6025-6030.

Daughton, C.G. and T.A. Ternes. 1999. Pharmaceuticals and personal care products in the environment: Agents of subtle change. Environ. Health. Persp. 107:907-944.

Fong, P.P., P.T. Huminski, and L.M. D'Urso. 1998. Induction and potentiation of parturition in fingernail clams (Sphaerium striatinum) by selective serotonin re-uptake inhibitors (SSRIs). J. Exp. Zool. 280:260-4.

Garcia-Reyero, N., E. Grau, M. Castillo, M.J. Lopez de Alda, D. Barcelo, and B. Pina. 2001. Monitoring of endocrine disruptors in surface waters by the yeast recombinant assay. Environ. Toxicol. Chem. 6:1152-8.

Gerwick, L., W.S. Reynolds, and C.J. Bayne. 2000. A precerebellin-like protein is part of the acute phase response in rainbow trout, Oncorhynchus mykiss. Dev. Comp. Immunol. 24:597-607.

Granjeaud, S., F. Bertucci, and B.R. Jordan. 1999. Expression profiling: DNA arrays in many guises. BioEssays. 21:781-790.

Graumann, K., A. Breithofer, and A. Jungbauer. 1999. Monitoring of estrogen mimics by a recombinant yeast assay: synergy between natural and synthetic compounds? Sci. Total. Environ. 225:69-79.

Halling-Sørenson, B., S. Nors Nielsen, P.F. Lanzky, F. Ingerslev, H.C. Holten Lützhøft, and S.E. Jergensen. 1998. Occurrence, fate and effects of pharmaceutical substances in the environment-a review. Chemosphere. 36:357-393.

Hjorth, S., H.J. Bengtsson, A. Kullberg, D. Carlzon, H. Peilot, and S.B. Auerbach. 2000. Serotonin autoreceptor function and antidepressant drug action. J. Psychopharmacol. 14:177-85.

Ishikawa, Y. 2000. Medakafish as a model system for vertebrate developmental genetics. BioEssays. 22:487-495.

Jorgensen, M., B. Vendelbo, N.E. Skakkebaek, and H. Leffers. 2000. Assaying estrogenicity by quantitating the expression levels of endogenous estrogen-regulated genes. Environ. Health. Perspect. 108:403-12.

Kersten, S., J. Seydoux, J.M. Peters, F.J. Gonzalez, B. Desvergne, and W. Wahli. 1999. Peroxisome proliferator-activated receptor alpha mediates the adaptive response to fasting. J. Clin. Invest. 103:1489-98

Kockx, M., P.P. Gervois, P. Poulain, B. Derudas, J.M. Peters, F.J. Gonzalez, H.M. Princen, T. Kooistra, and B. Staels. 1999. Fibrates suppress fibrinogen gene expression in rodents via activation of the peroxisome proliferator-activated receptor-alpha. Blood. 93:2991-8.

Koger, C.S., S.J. The, and D.E. Hinton. 1999. Variations of light and temperature regimes and resulting effects on reproduction parameters in medaka (*Oryzias latipes*). Biol. Repro. 61:1287-1293.

Leone, T.C., C.J. Weinheimer, and D.P. Kelly. 1999. A critical role for the peroxisome proliferator-activated receptor alpha (PPARalpha) in the cellular fasting response: the PPARalpha-null mouse as a model of fatty acid oxidation disorders. Proc. Natl. Acad. Sci. U S A. 96:7473-8.

Matthews, J., T. Celius, R. Halgren, and T. Zacharewski. 2000. Differential estrogen receptor binding of estrogenic substances: a species comparison. J. Steroid. Biochem. Mol. Biol. 74:223-34.

Mayer, F.L., D.J. Versteeg, M.J. McKee, L.C. Folmar, R.L. Graney, D.C. McCume, and B.A. Rattner. 1992. Physiological and nonspecific biomarkers. *In* R. J. Huggett, R. a. Kimerle, P.M. Mehrle, and H.L. Bergman (ed.), Biomarkers: Biochemic al., Physiologic al., and Histological Markers of Anthropogenic Stress. Lewis Publishers, Boca Raton, FL.

Medlin, J.F. 1999. Timely toxicology. Environ. Health. Perspect. 107:A256-A258.

Muschamp, J.W. and Fong, P.P. 2001. Effects of the serotonin receptor ligand methiothepin on reproductive behavior of the freshwater snail Biomphalaria glabrata: reduction of egg laying and induction of penile erection. J. Exp. Zool. 289:202-7.

Nuwaysir, E.F., M. Bittner, J. Trent, C. Barrett, and C. Afshari. 1999. Microarrays and toxicology: The advent of toxicogenomics. Mol. Carcinog. 24:153-159.

Papoulias, D.M., D.E. Tillitt, S. Jones, and D. Noltie. 1995. Use of medaka (*Oryzias latipes*) as a model to identify reproductive endpoints indicative of exposure to endocrine disrupting chemicals. Abstracts, 16th Annual Meeting, Society of Environmental Toxicology and Chemistry, Vancouver, BC, Canada, November 5-9. .

Patyna, P.J., R.A. Davis, T.F. Parkerton, R.P. Brown, and K.R. Cooper. 1999. A proposed multi-generation protocol for Japanese medaka (*Oryzias latipes*) to evaluate effects of endocrine disruptors. Sci. Total. Environ. 233:423-429.

Pennie, W.D., J.D. Tugwood, G.J. Oliver, and I. Kimber. 2000. The principles and practice of toxicogenomics: Applications and opportunities. Toxicol. Sci. 54:277-283.

Pharma.org. http://phrma.org/publications/publications/profile01/chapter2.phtml

Richardson, M.L. and J.M. Bowron, 1985. Fate of pharmaceutical chemicals in the environment-A review. J. Pharm. Pharacol. 37:1-12.

Rockett, J.C. and D.J. Dix. 2000. DNA arrays: technology, options and toxicological applications. Xenobiotica. 30:155-177.

Soto A.M., C. Sonnenschein, K.L. Chung, M.F. Fernandez, N. Olea, and F.O. Serrano. 1995. The E-SCREEN assay as a tool to identify estrogens: an update on estrogenic environmental pollutants. Environ. Health. Perspect. 103 Suppl 7:113-22.

Sedlak, D. 2000. Challenges associated with quantification of trace concentrations of pharmaceuticals in a complex matrix. Endocrine disruptors and pharmaceutical active compounds in drinking water. workshop. The University of Iowa Center for Health Effects of Environmental Contamination (CHEEC) April 19-21, 2000 in Chicago. ILO http://www.cheec.uiowa.edu/edc_2000/

Ternes, T.A. 1998. Occurrence of drugs in German sewage treatment plants and rivers. Wat. Res. 32:3245.

U.S. EPA. 1993a. Methods for Aquatic Toxicity Identification Evaluations: Phase II Toxicity Identification Procedures for Samples Exhibiting Acute and Chronic Toxicity. Office of Research and Development. Washington, D.C. EPA/600/R-92-080.

U.S. EPA. 1993b. Methods for Aquatic Toxicity Identification Evaluations: Phase III Toxicity Identification Procedures for Acutely and Chronically Toxic Samples. U.S. Environmental Protection Agency, Duluth, MN. EPA/600/R-92-081.

U.S. EPA. 1991. Technical Support Document for Water Quality-based Toxics Control. Office of Water. Washington, D.C. EPA/505/2-90-001.

U.S. EPA. 2000. Understanding and accounting for method variability in whole effluent toxicity applications under the National Pollutant Discharge Elimination System Program. D.L. Denton, J. Fox, F.A. Fulk, K. Greenwald, M. Narvaez, T.J. Norberg-King, L. Phillips. eds. EPA/833/R-00-003. Office of Water. Washington, D.C.

U.S. FDA. Food Quality Protection Act of 1996: report together with additional views (to accompany H.R. 1627) (including cost estimate of the Congressional Budget Office). [Washington, D.C.: U.S. G.P.O., 1996. Series title: Rept. / 104th Congress, 2d session, House of Representatives; 104-669.

Villalobos, A.S., M.J. Anderson, M.S. Denison, D.E. Hinton, K. Tullis, I.M. Kennedy, A.D. Jones, D.Y. Chang, G. Yang, and P. Kelly. 1996. Dioxinlike properties of a trichloroethylene combustion-generated aerosol. Environ. Health. Perspect. 104:734-743.

Willson, T.M. and W. Wahli. 1997. Peroxisome proliferator-activated receptor agonists. Curr. Opin. Chem. Biol. 1:235-41.

Yadetie, F., A. Laegreid, I. Bakke, W. Kusnierczyk, J. Komorowski, H.L. Waldum, and A.K. Sandvik. 2003. Liver gene expression in rats in response to the peroxisome proliferator-activated receptor-alpha agonist ciprofibrate. Physiol. Genomics. 15:9-19.

Yamazaki, K, J. Kuromitsu, and I. Tanaka. 2001. Microarray analysis of gene expression changes in mouse liver induced by peroxisome proliferator- activated receptor alpha agonists. Biochem. Biophys. Res. Commun. 290:1114-22.

Zwiener, C. and F.H. Frimmel. 2000. Oxidative treatment of pharmaceuticals in water. J. Chromatogr. A. 862:137-45.

WERF SUBSCRIBERS

WASTEWATER UTILITY

Alabama
Montgomery Water Works & Sanitary Sewer Board

Alaska
Anchorage Water & Wastewater Utility

Arizona
Glendale, City of, Utilities Department
Mesa, City of
Peoria, City of
Phoenix Water Services Department
Pima County Wastewater Management
Safford, City of

Arkansas
Little Rock Wastewater Utility

California
Calaveras County Water District
Central Contra Costa Sanitary District
Corona, City of
Crestline Sanitation District
Delta Diablo Sanitation District
Dublin San Ramon Services District
East Bay Dischargers Authority
East Bay Municipal Utility District
Eastern Municipal Water District
El Dorado Irrigation District
Fairfield-Suisun Sewer District
Fresno Department of Public Utilities
Inland Empire Utilities Agency
Irvine Ranch Water District
Las Virgenes Municipal Water District
Livermore, City of
Los Angeles, City of
Los Angeles County, Sanitation Districts of
Napa Sanitation District
Orange County Sanitation District
Palo Alto, City of
Riverside, City of
Sacramento Regional County Sanitation District
San Diego Metropolitan Wastewater Department, City of
San Francisco, City & County of
San Jose, City of
Santa Barbara, City of
Santa Cruz, City of
Santa Rosa, City of
South Bayside System Authority
South Coast Water District
South Orange County Wastewater Authority
Stege Sanitary District
Sunnyvale, City of
Union Sanitary District
West Valley Sanitation District

Colorado
Aurora, City of
Boulder, City of
Greeley, City of
Littleton/Englewood Water Pollution Control Plant
Metro Wastewater Reclamation District, Denver

Connecticut
Greater New Haven WPCA

District of Columbia
District of Columbia Water & Sewer Authority

Florida
Broward, County of
Fort Lauderdale, City of
Miami-Dade Water & Sewer Authority
Orange County Utilities Department
Reedy Creek Improvement District
Seminole County Environmental Services
St. Petersburg, City of
Tallahassee, City of
Tampa, City of
Toho Water Authority
West Palm Beach, City of

Georgia
Atlanta Department of Watershed Management
Augusta, City of
Clayton County Water Authority
Cobb County Water System
Columbus Water Works
Fulton County
Gwinnett County Department of Public Utilities
Savannah, City of

Hawaii
Honolulu, City & County of

Idaho
Boise, City of

Illinois
American Bottoms Wastewater Treatment Plant
Greater Peoria Sanitary District
Kankakee River Metropolitan Agency
Metropolitan Water Reclamation District of Greater Chicago
Wheaton Sanitary District

Iowa
Ames, City of
Cedar Rapids Wastewater Facility
Des Moines, City of
Iowa City

Kansas
Johnson County Unified Wastewater Districts
Unified Government of Wyandotte County/ Kansas City, City of

Kentucky
Louisville & Jefferson County Metropolitan Sewer District
Sanitation District No. 1

Louisiana
Sewerage & Water Board of New Orleans

Maine
Bangor, City of
Portland Water District

Maryland
Anne Arundel County Bureau of Utility Operations
Howard County Department of Public Works
Washington Suburban Sanitary Commission

Massachusetts
Boston Water & Sewer Commission
Upper Blackstone Water Pollution Abatement District

Michigan
Ann Arbor, City of
Detroit, City of
Holland Board of Public Works
Saginaw, City of
Wayne County Department of Environment
Wyoming, City of

Minnesota
Rochester, City of
Western Lake Superior Sanitary District

Missouri
Independence, City of
Kansas City Missouri Water Services Department
Little Blue Valley Sewer District
Metropolitan St. Louis Sewer District

Nebraska
Lincoln Wastewater System

Nevada
Henderson, City of
Reno, City of

New Jersey
Bergen County Utilities Authority
Ocean, County of
Passaic Valley Sewerage Commissioners

New York
New York City Department of Environmental Protection

North Carolina
Charlotte/Mecklenburg Utilities
Durham, City of
Metropolitan Sewerage District of Buncombe County
Orange Water & Sewer Authority

Ohio
Akron, City of
Butler County Department of Environmental Services
Columbus, City of
Metropolitan Sewer District of Greater Cincinnati
Northeast Ohio Regional Sewer District
Summit, County of

Oklahoma
Oklahoma City Water & Wastewater Utility Department
Tulsa, City of

Oregon
Clean Water Services
Eugene, City of
Gresham, City of
Portland, City of
Water Environment Services

Pennsylvania
Philadelphia, City of
University Area Joint Authority

South Carolina
Charleston Water System
Mount Pleasant Waterworks & Sewer Commission
Spartanburg Water

Tennessee
Cleveland, City of
Knoxville Utilities Board
Murfreesboro Water & Sewer Department
Nashville Metro Water Services

Texas
Austin, City of
Dallas Water Utilities
Denton, City of
El Paso Water Utilities
Fort Worth, City of
Houston, City of
San Antonio Water System
Trinity River Authority

Utah
Salt Lake City Corporation

Virginia
Alexandria Sanitation Authority
Arlington, County of

Fairfax County
Hampton Roads Sanitation District
Henrico, County of
Hopewell Regional Wastewater Treatment Facility
Loudoun County Sanitation Authority
Lynchburg Regional WWTP
Prince William County Service Authority
Richmond, City of
Rivanna Water & Sewer Authority

Washington
Everett, City of
King County Department of Natural Resources
Seattle Public Utilities
Sunnyside, Port of
Yakima, City of

Wisconsin
Green Bay Metro Sewerage District
Kenosha Water Utility
Madison Metropolitan Sewerage District
Milwaukee Metropolitan Sewerage District
Racine, City of
Sheboygan Regional Wastewater Treatment
Wausau Water Works

Australia
South Australian Water Corporation
Sydney Water Corporation
Water Corporation of Western Australia

Canada
Regina, City of, Saskatchewan
Toronto, City of, Ontario
Winnipeg, City of, Manitoba

New Zealand
Watercare Services Limited

United Kingdom
Yorkshire Water Services Limited

STORMWATER UTILITY

California
Fresno Metropolitan Flood Control District
Los Angeles, City of, Department of Public Works
Monterey, City of
Sacramento, County of
San Francisco, City & County of
Santa Rosa, City of
Sunnyvale, City of

Colorado
Aurora, City of
Boulder, City of

Georgia
Griffin, City of

Iowa
Cedar Rapids Wastewater Facility
Des Moines, City of

Kansas
Overland Park, City of

Kentucky
Louisville & Jefferson County Metropolitan Sewer District

Maine
Portland Water District

North Carolina
Charlotte, City of, Stormwater Services

Pennsylvania
Philadelphia, City of

Tennessee
Chattanooga Stormwater Management

Texas
Harris County Flood Control District, Texas

Washington
Bellevue Utilities Department
Seattle Public Utilities

STATE

Arkansas Department of Environmental Quality
Connecticut Department of Environmental Protection
Kansas Department of Health & Environment
Kentucky Department of Environmental Protection
Ohio River Valley Sanitation Commission
Urban Drainage & Flood Control District, CO

CORPORATE

ADS Environmental Services
Advanced Data Mining International
Alan Plummer & Associates
Alden Research Laboratory Inc.
Alpine Technology Inc.
Aqua-Aerobic Systems Inc.
Aquateam–Norwegian Water Technology Centre A/S
ARCADIS
Associated Engineering
Black & Veatch
Boyle Engineering Corporation
Brown & Caldwell
Burgess & Niple, Ltd.
Burns & McDonnell
CABE Associates Inc.
The Cadmus Group
Camp Dresser & McKee Inc.
Carollo Engineers Inc.
Carpenter Environmental Associates Inc.
CDS Technologies Inc.
CET Engineering Services
Chemtrac Systems Inc.
CH2M HILL
Conestoga-Rovers & Associates
CONTECH Stormwater Solutions
D&B/Guarino Engineers, LLC
Damon S. Williams Associates, LLC
Earth Tech Inc.
Ecovation
EMA Inc.
Environ/The ADVENT Group, Inc.
Fay, Spofford, & Thorndike Inc.
Freese & Nichols Inc.
ftn Associates Inc.
Gannett Fleming Inc.
Garden & Associates, Ltd.
Geosyntec Consultants
GHD
Golder Associates Ltd.
Greeley and Hansen LLC
Hazen & Sawyer, P.C.
HDR Engineering Inc.
HNTB Corporation
Hydromantis Inc.
HydroQual Inc.
Infilco Degremont Inc.
Jacobson Helgoth Consultants Inc.
Jason Consultants LLC Inc.
Jordan, Jones, & Goulding Inc.
KCI Technologies Inc.
Kelly & Weaver, P.C.
Kennedy/Jenks Consultants
KMK Consultants
Komline Sanderson Engineering Corporation
Limno-Tech Inc.
Lombardo Associates Inc.
The Low Impact Development Center Inc.
Malcolm Pirnie Inc.
Material Matters
McKim & Creed
Metcalf & Eddy Inc.
Monteco Corporation
MPR Engineering Corporation, Inc.
MWH
NewFields Water Resources, LLC
O'Brien & Gere Engineers Inc.
Odor & Corrosion Technology Consultants Inc.
Oscar Larson & Associates
PA Government Services Inc.
Parametrix Inc.
Parsons
Post, Buckley, Schuh & Jernigan
Praxair, Inc.
RMC Water & Environment
Ross & Associates Ltd.
Rothberg, Tamburini & Windsor, Inc.
SAIC
Savin Engineers
Siemens Water Technologies
Stantec Consulting Inc.
Stearns & Wheler, LLC
Stone Environmental Inc.
Stratus Consulting Inc.
Synagro Technologies Inc.
Tetra Tech Inc.
Trojan Technologies Inc.
Trussell Technologies, Inc.
URS Corporation
Wade-Trim Inc.
Westin Engineering Inc.
Weston Solutions Inc.
Woodard & Curran
Zenon Environmental Inc.
Zoeller Pump Company

INDUSTRY

American Electric Power
American Water
ChevronTexaco Energy Research & Technology Company
The Coca-Cola Company
Dow Chemical Company
DuPont Company
Eastman Chemical Company
Eli Lilly & Company
Merck & Company Inc.
Premier Chemicals LLC
Procter & Gamble Company
Thames Water Utilities
Severn Trent Services Inc.
Suez Environnment
United Water Services LLC

Note: List as of 2/1/07

WERF Board of Directors

WERF Research Council

WERF Product Order Form

As a benefit of joining the Water Environment Research Foundation, subscribers are entitled to receive one complimentary copy of all final reports and other products.Additional copies are available at cost (usually $10).To order your complimentary copy of a report, please write "free" in the unit price column.WERF keeps track of all orders.If the charge differs from what is shown here, we will call to confirm the total before processing.

Name ______________________ Title ______________________

Organization ______________________

Address ______________________

City ________ State ________ Zip Code ________ Country ________

Phone ________ Fax ________ Email ________

Stock #	Product	Quantity	Unit Price	Total
			Postage & Handling	
			VA Residents Add 4.5% Sales Tax	
			Canadian Residents Add 7% GST	
			TOTAL	

Method of Payment:

(All orders must be prepaid.)

❑ Check or Money Order Enclosed

❑ Visa ❑ Mastercard ❑ American Express

Account No. ______________________ Exp. Date ________

Signature ______________________

Shipping & Handling:

Amount of Order	United States	Canada & Mexico	All Others
Up to but not more than:	Add:	Add:	Add:
$20.00	$5.00*	$8.00	50% of amount
30.00	5.50	8.00	40% of amount
40.00	6.00	8.00	
50.00	6.50	14.00	
60.00	7.00	14.00	
80.00	8.00	14.00	
100.00	10.00	21.00	
150.00	12.50	28.00	
200.00	15.00	35.00	
More than $200.00	Add 20% of order	Add 20% of order	

*minimum amount for all orders

Note: Please make checks payable to the Water Environment Research Foundation.

To Order (Subscribers Only):

Log on to www.werf.org and click on the "Product Catalog."

Phone: (703) 684-2470
Fax: (703) 299-0742.

WERF
Attn: Subscriber Services
635 Slaters Lane
Alexandria, VA 22314-1177

To Order (Non-Subscribers):

Non-subscribers may be able to order WERF publications either through WEF (www.wef.org) or IWAP (www.iwapublishing.com).Visit WERF's website at www.werf.org for details.

IWA Publishing's authorised EU representative for General Product Safety Regulations is Diane D'Arras, 15 rue Duret, 75116 Paris, France, e-mail: safety@iwap.co.uk.

Printed and bound by CPI Group (UK) Ltd, Croydon, CR0 4YY
12/05/2026
02108874-0009